Lecture Notes in Computer Science 13773

More information about this series at https://link.springer.com/bookseries/558

Partha Pratim Roy · Arvind Agarwal · Tianrui Li ·
P. Krishna Reddy · R. Uday Kiran (Eds.)

Big Data Analytics

10th International Conference, BDA 2022
Hyderabad, India, December 19–22, 2022
Proceedings

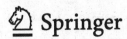 Springer

Editors
Partha Pratim Roy ⓘ
Indian Institute of Technology-Roorkee
Roorkee, India

Arvind Agarwal ⓘ
IBM Research
Gurugram, India

Tianrui Li ⓘ
Southwest Jiaotong University
Chengdu, China

P. Krishna Reddy ⓘ
International Institute of Information
Technology - Hyderabad
Hyderabad, India

R. Uday Kiran ⓘ
The University of Aizu
Fukushima, Japan

ISSN 0302-9743   ISSN 1611-3349 (electronic)
Lecture Notes in Computer Science
ISBN 978-3-031-24093-5   ISBN 978-3-031-24094-2 (eBook)
https://doi.org/10.1007/978-3-031-24094-2

This Springer imprint is published by the registered company Springer Nature Switzerland AG
The registered company address is: Gewerbestrasse 11, 6330 Cham, Switzerland

Preface

Many real-world applications naturally produce big data. This data hides hidden information that can empower the end-users to gain competitive insights and achieve socio-economic development. However, finding this helpful information is non-trivial and challenging due to the 6V's (volume, variety, velocity, veracity, value, and variability) nature of big data. Therefore, we need to explore innovative analytical solutions by drawing expertise from recent advances in several fields, such as data mining, database systems, statistics, distributed data processing, machine/deep learning, and artificial intelligence. There is also a vital need to develop sustainable and resilient data analysis systems for emerging applications such as autonomous vehicles, blockchains, healthcare, and IoT.

The 10th International Conference on Big Data Analytics (BDA 2022) was held online and offline during December 19–22, 2022 at the International Institute of Information Technology, Hyderabad, India. The proceedings includes 14 peer-reviewed research papers and four contributions by distinguished keynote and invited speakers. This year's program covered a wide range of topics related to big data analytics, such as data science, pattern mining, graph analytics, machine learning, healthcare analytics, information extraction, and data analytics in agriculture.

The research papers, keynote speeches, invited talks, and tutorials presented at the conference will promote research in big data analytics and stimulate the development of innovative solutions to industry and societal problems.

The conference received 36 submissions. The review process of research papers followed a single-blind two-tiered review system following the tradition of BDA. We employed Microsoft's Conference Management Toolkit (CMT) for the paper submission and review processes. The Program Committee (PC) consisted of competent data analytical researchers from academia and industry based in several countries. Each submission was reviewed by at least two to three reviewers from the Program Committee (PC) and discussed by the PC chairs before a decision was made. Based on the above review process, the Program Committee accepted seven full papers (acceptance ratio of 17%) and seven short papers (acceptance ratio of 17%).

The BDA 2022 program included four keynote talks, three invited talks, 14 research papers, a panel, four tutorials, and four workshops. The four keynote talks were delivered by Y. Narahari (Indian Institute of Science, Bangalore, India), Raj Sharman (University at Buffalo, USA), Philippe Fournier-Viger (Shenzhen University, China), and Sanjay Madria (Missouri University of Science and Technology, USA). The three invited talks were delivered by S. Bapi Raju (International Institute of Information Technology-Hyderabad, Telangana, India), Arvind Agarwal (IBM-Research), and Sridhar Viswanathan (Bank of America).

Four workshops were selected by the workshop co-chairs to be held in conjunction with BDA 2022: Data Challenges in Assessing (Urban & Regional) Air Quality (DACAAQ 2022), Big Data Analytics on High-Performance Computing Cluster (HPCC)

Systems, the Workshop on Universal Acceptance and Email Address, and the Workshop on Data Science for Justice Delivery in India (DSJDI 2022).

We want to thank all PC members for their hard work in providing us with thoughtful and comprehensive reviews and recommendations. Many thanks to the authors who submitted their papers to the conference. The Steering Committee and the Organizing Committee deserve praise for their support. Several volunteers contributed to the success of the conference. We thank all keynote speakers, invited speakers, workshop speakers, and panelists for their insightful talks. Lastly, we also acknowledge the International Institute of Information Technology-Hyderabad for hosting the conference. IIIT Hyderabad is celebrating its silver jubilee in 2022. We thank iHub-Data, IIIT-Hyderabad, for sponsoring the conference. We also thank all partners for supporting the conference.

We hope that the readers of the proceedings find the content interesting, rewarding, and beneficial to their research.

December 2022

Partha Pratim Roy
Arvind Agarwal
Tianrui Li
P. Krishna Reddy
R. Uday Kiran

Organization

Honorary Chairs

P. J. Narayanan   IIIT Hyderabad, India
B. Srinivasa Murty   IIT Hyderabad, India
Sharma Chakravarthy   The University of Texas at Arlington, USA

General Chairs

P. Krishna Reddy   IIIT Hyderabad, India

Steering Committee Chair

P. Krishna Reddy   IIIT Hyderabad, India

Steering Committee Members

Srinath Srinivasa   IIIT Bangalore, India
Mukesh Mohania   IIIT Delhi, India
Sharma Chakravarthy   The University of Texas at Arlington, USA
Philippe Fournier-Viger   Shenzhen University, China
Sanjay Kumar Madria   Missouri University of Science and Technology, USA
Masaru Kitsuregawa   University of Tokyo, Japan
Raj K. Bhatnagar   University of Cincinnati, USA
Vasudha Bhatnagar   University of Delhi, India
Ladjel Bellatreche   ISAE-ENSMA, France
H. V. Jagadish   University of Michigan, USA
Ramesh Kumar Agrawal   Jawaharlal Nehru University, India
Divyakant Agrawal   University of California at Santa Barbara, USA
Arun Agarwal   University of Hyderabad, India
Jaideep Srivastava   University of Minnesota, USA
Anirban Mondal   Ashoka University, India
Sanjay Chaudhary   Ahmedabad University, India
Jerry Chun-Wei Lin   Western Norway University of Applied Sciences, Norway
S. K. Gupta   IIIT Delhi, India
Krithi Ramamritham   IIT Bombay, India

Program Committee Chairs

Partha Pratim Roy	IIT Roorkee, India
Arvind Agarwal	IBM Research, India
Tianrui Li	Southwest Jiaotong University, China

Publication Chair

R. Uday Kiran	The University of Aizu, Japan

Sponsorship Chair

Anirban Mondal	Ashoka University, India

Workshop Chairs

Satish Srirama	University of Hyderabad, India
P. Radha Krishna	NIT Warangal, India
Sonali Agarwal	IIIT Allahabad, India

Tutorial Chairs

Vikram Goyal	IIIT Delhi, India
Rajeev Gupta	Microsoft, India
Y. Kalidas	IIT Tirupati, India

Publicity Chairs

Tai Dinh	The Kyoto College of Graduate Studies for Informatics, Japan
M. Kumara Swamy	CMR Engineering College, Hyderabad, India
P. Sateesh Kumar	IIT Roorkee, India

Panel Chairs

Vikram Goyal	IIIT Delhi, India
Rajeev Gupta	Microsoft, India
Y. Kalidas	IIT Tirupati, India

Organizing Committee

A. Srinivas Reddy	IIIT Hyderabad, India
U. Narendra Babu	IIIT Hyderabad, India
A. Mamatha	IIIT Hyderabad, India
N. Chandra Shekar	IIIT Hyderabad, India
C. Saideep	IIIT Hyderabad, India

Program Committee

Amit Singh	Indian Institute of Technology, Dhanbad, India
Anjali Gautam	Indian Institute of Technology Roorkee, India
Chinmaya Dehury	University of Tartu, Estonia
Chongshou Li	Southwest Jiaotong University, China
Chuan Luo	Sichuan University, China
Divya Sardana	IEEE Member
Engelbert Mephu Nguifo	University of Clermont Auvergne, CNRS, LIMOS, France
Fengmao Lv	Southwest Jiaotong University, China
Gautam Srivastava	Brandon University, Canada
Jerry Chun-Wei Lin	Western Norway University of Applied Sciences, Norway
Ko-Wei Huang	National Kaohsiung University of Science and Technology, Taiwan
Philippe Fournier-Viger	Shenzhen University, China
Pradeep Kumar	IIT Roorkee, India
Prashant Srivastava	University of Allahabad, India
Raj Bhatnagar	University of Cincinnati, USA
Rajkumar Saini	Luleå Tekniska Universitet, Sweden
Ritwik Chaudhuri	IBM Research, India
Rohith D. Vallam	IBM Research, India
Sanjay Chaudhary	Ahmedabad University, India
Sebastián Ventura Soto	University of Cordoba, Spain
Shih Hsiung Lee	National Kaohsiung University of Applied Sciences, Taiwan
Stefania Tomasiello	University of Tartu, Estonia
Thippa Reddy Gadekallu	Vellore Institute of Technology, India
Uday Rage	University of Aizu, Japan
Vasudha Bhatnagar	University of Delhi, India
Vikram Singh	National Institute of Technology, Kurukshetra, India
Vishesh Tanwar	University of Central Florida, USA
Vitobha Munigala	IBM Research AI, India
Xin Yang	Southwestern University of Finance and Economics, China

Contents

Big Data Analytics: Vision and Perspectives

Data Challenges and Societal Impacts – The Case in Favor of the Blueprint
for an AI Bill of Rights (Keynote Remarks) 3
   Raj Sharman

Big Data in Cognitive Neuroscience: Opportunities and Challenges 16
   Kamalaker Dadi and Bapi Raju Surampudi

Data Science: Architectures

A Novel Feature Selection Based Text Classification Using Multi-layer
ELM 33
   Rajendra Kumar Roul and Gaurav Satyanath

ARCORE: A Requirements Dataset for Service Identification 53
   Vijaya Peketi and Surekha Satti

Learning Enhancement Using Question-Answer Generation for e-Book
Using Contrastive Fine-Tuned T5 68
   Shobhan Kumar, Arun Chauhan, and Pavan Kumar C.

Data Science: Applications

A Machine and Deep Learning Framework to Retain Customers Based
on Their Lifetime Value 91
   Kannan Kumaran, Pramod Pathak, Rejwanul Haque, and Paul Stynes

A Deep Learning Based Approach to Automate Clinical Coding
of Electronic Health Records 104
   Ashutosh Kumar and Santosh Singh Rathore

Determining the Severity of Dementia Using Ensemble Learning 117
   Shruti Srivatsan, Sumneet Kaur Bamrah, and K. S. Gayathri

A Distributed Ensemble Machine Learning Technique for Emotion
Classification from Vocal Cues 136
   Bineetha Vijayan, Gayathri Soman, M. V. Vivek, and M. V. Judy

Graph Analytics

Drugomics: Knowledge Graph & AI to Construct Physicians' Brain
Digital Twin to Prevent Drug Side-Effects and Patient Harm 149
    Asoke K. Talukder, Erwin Selg, Ryan Fernandez, Tony D. S. Raj,
    Abijeet V. Waghmare, and Roland E. Haas

Extremely Randomized Tree Based Sentiment Polarity Classification
on Online Product Reviews 159
    R. B. Saranya, Ramesh Kesavan, and K. Nisha Devi

Community Detection in Large Directed Graphs 172
    Siqi Chen and Raj Bhatnagar

Pattern Mining

FastTIRP: Efficient Discovery of Time-Interval Related Patterns 185
    Philippe Fournier-Viger, Yuechun Li, M. Saqib Nawaz, and Yulin He

Discovering Top-k Periodic-Frequent Patterns in Very Large Temporal
Databases 200
    Palla Likhitha, Penugonda Ravikumar, Rage Uday Kiran,
    and Yutaka Watanobe

Hui2Vec: Learning Transaction Embedding Through High Utility Itemsets 211
    Khaled Belghith, Philippe Fournier-Viger, and Jassem Jawadi

Predictive Analytics in Agriculture

A Data-Driven, Farmer-Oriented Agricultural Crop Recommendation
Engine (ACRE) 227
    Rohit Patel, Inavamsi Enaganti, Mayank Ratan Bhardwaj,
    and Y. Narahari

Analyze the Impact of Weather Parameters for Crop Yield Prediction
Using Deep Learning 249
    Pragneshkumar Patel, Sanjay Chaudhary, and Hasit Parmar

Analysis of Weather Condition Based Reuse Among Agromet Advisory:
A Validation Study 260
    Mamatha Alugubelly, Krishna Reddy Polepalli, Anirban Mondal,
    S. G. Mahadevappa, Balaji Naik Banoth, and Sreenivas Gade

Author Index 279

Big Data Analytics: Vision and Perspectives

Data Challenges and Societal Impacts – The Case in Favor of the Blueprint for an AI Bill of Rights (Keynote Remarks)

Raj Sharman[✉]

University at Buffalo, Buffalo 14214, USA
rsharman@buffalo.edu

Abstract. Artificial Intelligence (AI) technologies contribute tremendously to various areas of life and society. Therefore, we are witnessing massive investments in this area. Grand-view Research has calibrated the global AI market at 93.5 billion dollars as of 2021. Further, according to their report, it is expected to grow at a compound annual growth rate (CAGR) of 38.1% from 2022 to 2030. It is believed that nations that adopt and use AI will have a competitive edge. AI, in some form, will be part of most products and services we use. This manuscript documents some of the challenges, tradeoffs, and remedies concerning BIG DATA in the age of Artificial Intelligence. The paper also provides a brief overview of the maladies that plague the landscape of BIG DATA and some academic literature that provides solutions to the problems in the context of AI. The driving motivation for writing this article is to highlight our responsibility to create algorithms and automated systems that do not harm and are equitable and just. I also hope to create awareness that leads to businesses and software laboratories that focus on testing software and data that alleviates our fear - modeling the work of the US Food and Drug Administration. The tenets echoed in this concord with the Blueprint for an AI Bill of Rights released by the United States (US) White House Office of Science and Technology Policy (OSTP) on October 4, 2022, and the AI Risk Management Framework by the US National Institute of Standards and Technology.

Keywords: Artificial intelligence · Big data · Intelligent systems · Machine learning · Deep leaning · Bill of rights · Autonomous systems

1 Introduction

At the outset, we explain a few developments in the area of BIG DATA and elucidate some traditional challenges with BIG DATA that still plague us. Gartner, a 4.7 billion dollar company as of 2022, with over 100 businesses in over 100 countries, on an annual basis, develops a highly read document for industry professionals entitled "The Gartner hype-cycle chart." This document highlights

P. P. Roy et al. (Eds.): BDA 2022 India, LNCS 13773, pp. 3–15, 2022.
https://doi.org/10.1007/978-3-031-24094-2_1

technology trends that classify emerging technologies into five stages: Innovation Trigger, Peak Inflated Expectations, Trough of Disillusionment, Slope of Enlightenment, and Plateau of Productivity. After several years of featuring as part of the Gartner Life Cycle, it has dropped the term 'BIG DATA' from its annual hype cycle. The reason is that it is indeed ubiquitous. BIG DATA is everywhere. The global BIG DATA analytics market is projected to be 271.83 billion by 2022 and 655.53 billion by 2029 [3]. Digitalization at an accelerated pace across retail, manufacturing, healthcare, and other industries is partially casual for the growth of the BIG DATA market. Further, vast amounts of data are generated from social media such as Tiktok, Facebook, Twitter, Instagram, Snapchat, and WhatsApp, and Online healthcare social networks such as TuDiabetes, PatientsLikeMe me, etc. Then there are tracking devices, such as smart phones, wearable gadgets, etc., contributing to these trends. This trend has led to a massive demand for data analytics to decipher information and personal or various purposes. A lot of the data and analysis is being done in real-time by automated systems using AI, often without the consent of the concerned people -resulting in the dawn of the age of Artificial Intelligence (AI). We discuss issues relating to AI in Sect. 2. However, at the outset, I allude to a few of the challenges, tradeoffs, and remedies concerning automated systems that employ BIG DATA still plague us. For the purposes of this manuscript, we use a more encompassing term, 'Automated Systems,' to include all applications using BIG DATA, including AI and Machine learning systems. Further, we assume that BIG DATA characterizes all data that may have, in addition to large volume, some of the following properties: Veracity, Velocity, and Variety. We list a few pertinent issues below (note this is not meant to be a comprehensive list):

- Volume Issue: Typically, when we use the term BIG DATA we realize that the volume of data is so massive that traditional data processing software may be unable to process them. For example, some applications process data in memory for performance reasons and therefore fail. To name one, the memory problems with Python are well documented. Memory storage issues in Python can lead to memory leaks preventing the program from working efficiently. That said, there may be workarounds to fix such problems.
- Veracity Issue: The issue of integrity is even more insidious. When data from one or more sources cannot be trusted for its authenticity/quality or the quality changes over time, the collection of the BIG DATA set can have integrity and other quality issues. When automated programs process this data, surely the results and predictions can harm people, generate spurious results (predictions), etc.
- Velocity Issue: When data collection occurs at a very rapid rate (high-velocity issue), traditional real-time analysis applications may not be able to process them in real-time. These are typical and well-understood challenges relating to BIG DATA. There are still other problems.
- Other
  - Often, machine learning algorithms learn from data collected from one or more sources/sensors continuously. The outputs of prediction algorithms

using real-time data then requires an infrastructure that can rapidly move large volumes of data across geographies. When a nation has inadequate infrastructure, BIG DAYA and AI programs may not work effectively. Therefore, we must all advocate for public or private solutions to address this issue fast so that, as a nation, we can all prosper from the benefits of BIG DATA and AI.

- Finally, the cost of storage, even with scaling (storage on the cloud), continues to pose challenges for small and medium-sized businesses, impacting the effective use of machine learning and AI algorithms in these organizations. These are but some traditional challenges. More severe issues related to harm undergird automated systems. Next (see Table 1), we discuss the harms categorized by the National Institute of Standards and Technology (NIST) [8].

Table 1. Types of harms

Harm to an Individual	This comes into play when a person's civil liberties or rights are jeopardized
Harm to a community or group	This type of harm arises when an automated system works in a manner that discriminates against a group. For example, the failure of the automated system to detect the fake account masquerading as a Twitter handle led to a billion-dollar loss. In addition, the Twitter account masquerading as the company Eli Lily handle informed the community that insulin would be provided free of cost. This led to millions selling their stocks.
Harm to an organization /Enterprise	The harm that can impact technical systems, organizational, assets, business operations and processes. For example, over-optimizing drive routes and intense scheduling may lead to employees being treated inhumanely,impacting employees (who are the organizational assets) in the long run.
Harm to a system	Harm to the interconnected system leads to cascading effects and multiple system failures.

2 Benefits of AI

The use of AI is rapidly becoming part of every facet of society. For example, it is being used in genomics, image and video processing, materials, natural language processing, robotics, wireless spectrum monitoring, predictions (i.e., predicting - recidivism, the chance of stroke, customer behavior, and more). This large-scale deployment and use profoundly affect our lives, culture, and society. To provide a more balanced perspective, it is essential to establish more concretely that AI has already proven beneficial to society in many ways. We describe three widely publicized (in the US media) use cases that show how AI has benefited society.

Case 1: Analyzing medical images such as MRI images, ultrasound results, cardiograms, and CT scan findings takes a lot of time. The images are graphical in nature, and further, the depth of each pixel is more than what the human eye can efficiently process. Sometimes the analysis with the naked eye can lead to missed signals. AI-based technologies automate medical documents' recording and help make a more accurate diagnosis, minimizing errors quickly and at a much lower cost. This ushers in excellent efficiency to the process allowing expert radiologists and other concerned physicians to focus on the more complicated cases. AI Algorithms process large volumes of data very quickly and trace patterns and relationships, allowing for quick and accurate diagnosis. Therefore, such systems have been widely adopted in many countries, contributing to the demand for such systems. The societal impacts of AI are tremendous. In addition, technology helps hospitals' bottom line and contributes to making these institutions more sustainable.

Case 2: An Israeli-based company is developing AI-based tools using deep learning software on CT scans to help better identify and target intracranial bleeding. The AI-based automated tools also help identify or rule out strokes or traumatic brain injuries in emergency rooms and ambulances. This impacts patients. Again, the societal impacts are tremendous.

Case 3: General Motors is investing in AI technologies for inspecting vehicles. Using sensors of various kinds, the automated system allows quick identification of defective parts. This improves passenger safety and, when fully deployed, will reduce costs. The contribution to productivity is significant for General Motors. But better and quicker diagnosis leads to a safer society.

With so many positive use cases of AI, it should be no surprise that AI contributes tremendously to various areas of life and society. Therefore, we are witnessing massive investments in this area. Grand-view Research AI Market Size Report, 2022–30 [2], has calibrated the global AI market at 93.5 billion as of 2021. Further, according to their report, it is expected to grow at a compound annual growth rate (CAGR) of 38.1% from 2022 to 2030. It is believed that nations that adopt and use AI will have a competitive edge. AI, in some form, will be part of most products and services we use. Developed nations with excellent broadband infrastructure will have an advantage in deploying and using AI more universally. The gap between developed and developing countries will only widen. Therefore, nations must quickly develop robust information infrastructure and Information highways to better harness the power of AI.

While we have progressed as a society to deploy AI in many walks of life, it is essential to work to mitigate the harm that AI can potentially cause. As in manufacturing, even with automated systems, it is better to engage in prevention first. This means understanding how machine learning systems are developed and deployed. We must bring to bear the understanding that AI and automated systems are technical systems that operate in a social system. Once AI systems are deployed, machines learn from the data it continues to collect from and about users. They may develop biases after they are deployed. Therefore there is a need for continuous monitoring and auditing systems - to detect and correct biases that may develop with use. The following section details biases that automated systems may suffer.

3   AI Bias

To begin this section, I present two use cases on the biases with automated systems that have been widely reported in the USA. We know 'Garbage In, Garbage Out' - a phrase popularized in the 1980s relating to IT systems. That adage still holds in the era of AI. AI, or Machine learning bias, occurs when an algorithm produces systemically prejudiced results due to erroneous assumptions in the machine learning process, program design, and data fed to the machine. Sometimes these biases are implicit and hard to detect. Sometimes they are more explicit. In the following two sections, this manuscript details a few of the significant issues relating to understanding biases and the subject of Trustworthiness. The manuscript's component builds on NIST Special Publications (1270) [9] and European Union (EU) [5].

Case 1: Amazon's e-commerce systems and automation have undoubtedly contributed to Amazon's market dominance. These automated systems are both functional and easy to use. Therefore, some systems were used without proper checks and balances regarding societal expectations. In 2015, researchers found that the automated system at Amazon, Inc. used for recruitment was gender-biased. As far as we know, there was no intention to develop a biased system. Amazon's machine learning models were trained on resumes submitted to the company over ten years. Most came from men. Amazon's system taught itself that male candidates were preferable. A section of society was denied the opportunity to work at Amazon. Further, possibly, some underqualified people were recruited. Amazon stopped using the machine learning program and disbanded the team (Dastin, 2018). When trained on data that includes human biases, historical inequalities, or different metrics of judgment based on gender, race, nationality, sexual orientation, etc., AI systems can cause great harm.

Case 2: Correctional Offender Management Profiling for Alternative Sanctions (COMPAS), an automated system, is used in the USA to predict which criminals are more likely to re-offend. Judges make decisions using the outcome of this AI system. Using COMPAS, black defendants were incorrectly labeled as "high-risk" and labeled as likely to commit a future crime twice as often as their white counterparts. The company that developed the software denied the claims made against it (Mesa, 2021). The more significant concern here is that algorithms ignore the social systems and structures in which they are created. In addition, the algorithms use biased data, so the predictions reinforce the stereotypical attitude. In short black people do not get relief from such automated systems.

Bias in AI can harm humans. For example, AI can make decisions that affect whether a person is admitted into schools and colleges, authorized for a bank loan, or accepted as a rental or loan applicant. In addition, it is common knowledge that AI systems can exhibit biases stemming from algorithms or data (used at the time of creation or added after use). The following section briefly clarifies the different biases that may infect automated systems. Biases may be classified [7] as (a) Statistical, (b) Human, and (c) Systemic. Further, biases may also be classified as implicit or explicit.

3.1 Statistical Bias

In this subsection, we provide a very brief introduction to the topic of Statistical Bias. This section is intentionally kept short though there is a lot of content to cover because many books on sampling, econometrics, and statistics provide more detailed explanations. Statistical biases manifest when there is a problem with experimental design or how the data is collected. When a statistic is not representative of the population, we consider this a statistical bias, possibly stemming from a sample selection, missing variables, etc. The analysis of biased samples leads to the actual parameters of a population being different from the statistics used to estimate the parameters (Harvard Business School Blog post, 2022 [11]). Statistical Bias includes Selection bias, Survivorship bias, Omitted variable Bias, Common-Method Bias, Protopathic Bias, etc.

Selection bias implies selective inclusion, which may or may not be by intent [16]. Survivorship bias is a type of selection bias when we do not consider the unobserved data [13]. When the dataset includes records relating only to success (customers/people who chose to partake in social media) or only defects (defective samples, customers who went to the competition, people who did not comment on a post) to inform the analysis, the results are lopsided. In some sense, this is a sort of sampling bias or measurement bias.

Omitted variable Bias is the difference between an estimator's expected value and the parameter's true value as a consequence of the model not including the appropriate predictor variables [15]. Econometrists typically deal with missing variables by finding instrument variables and engaging in two-stage least square regression. That said, there are other remedies also available to developers.

Protopathic Bias occurs when the applied treatment for a disease or outcome appears to cause the outcome [12]. Protopathic Bias occurs when the drug is initiated in response to the first symptoms of the disease, which is, at this point, undiagnosed. The results may lead to the interpretation that the drug caused the disease (i.e., the results are misinterpreted as a reverse causality effect) [14].

A thorough and rigorous analysis will address all such identification issues.

3.2 Human and Systemic Bias

A complete understanding of Bias must consider human and systemic biases. Human Bias occurs when an automated system or algorithm produces systemically prejudiced results due to erroneous assumptions in the machine learning process. Systemic biases result from institutions operating in ways that disadvantage certain social groups, leading to discrimination against individuals. Human biases can relate to how people interpret and use data to fill in missing information, such as a person's neighborhood of residence influencing how likely authorities would consider the person a crime suspect. These biases also develop because of how data is collected and processed, leading to biased conclusions. More accurate algorithms may be unable to fix issues using opportunistically collected data. A solution to build fairer algorithms trained on more representative data needs to consider equality and equity. Fairer data may need to be synthetically created if one is not available using previously collected data. This new data can then be combined with existing data in judicious ways. But then, how do we know that the training data is correct?

3.3 Beyond Bias - The Issue of Consent

Beyond bias, nations may have to deal with the issue of consent through legislation. A lot of data, including the digital footprints of people who use automated systems (such as tracking systems, phone apps, and the internet in general), are being harnessed without the express consent of the users. The purposes for such data collection vary but often, the explanation provided is 'to provide better service to customers. This collection has led to two trends: (a) commodification of their digital footprint without proper consent and (b) processing of data using automated systems such as AI (machine learning algorithms) without regard to the potential harm it may create. The question is: Should we allow this to continue? Do people have rights and protections from those collecting data without consent? Without fundamental digital rights, information in the information super highway is akin to the Wild West of yesteryear. Governments across the globe need to think of legislation that demands transparency when automated systems are used. True implementing these safeguards will come at a cost to the industry (which may be passed on to customers). Governments need to develop legislations that require those using automated systems to assure their citizens that these systems are not biased. When there is information asymmetry, the hypocrites' oath of "Do no harm" helps when system developers embody it. We need to create a new generation of programmers and system developers who will embody this principle.

4 A Framework to Build Trustworthiness

To protect citizens from potential horrors stemming from the unchecked use of AI, many countries are developing national AI strategies and policies that inform governments and the industry about the responsible use of technology. In April 2021, the EU Commission presented a coordinated plan for Artificial Intelligence. More details on this can be found on the EU [4] website on AI. An EU proposal for legislation addresses the risks of specific uses of AI, categorizing them into four levels: unacceptable risk, high risk, limited risk, and minimal risk. In the United States, the National Institute of Standards and Technology (NIST) is developing a framework to better manage risks to individuals, organizations, and society associated with artificial intelligence (AI). This framework is helpful for programmers, developers, system designers, and business managers to consider while designing, deploying, and testing their automated systems. The framework also recognizes that once deployed, the AI systems learn from users and also from other data available to them to make predictions that lead to decisions. Therefore, safeguards and audit procedures need to be built to monitor AI systems on an ongoing basis to ensure that biases are not introduced. Parameters influencing the outcome of AI systems need adjusting as new legislation or policies are created – this entails feeding the system sufficient new data to function in the new environment (functioning of algorithms after new legislation or policy develops). Therefore, it is crucial that self-learning machines are better governed. There is a need for continuous monitoring and audit.

Since the proposed framework by NIST provides new directions for improving Trustworthiness, we provide a brief of the salient aspects. The proposed framework, for the most part, addresses the following two main issues: (a) Technical Characteristics and (b) Socio-Technical Characteristics

4.1   Technical Characteristics

Technical characteristics address discussed here include (a) Accuracy, (b) Reliability, (c) Robustness, and (d) Resilience and Security. It may be noted that this subsection borrows content from the NIST AI framework proposal, as I advocate for incorporating its tenets as we develop systems and Software. We provide a brief on each of these characteristics in the following subsections:

Accuracy: Accuracy indicates the degree to which the ML model correctly captures a relationship within training data. AI risk management processes should consider the potential risks to the enterprise and society if the underlying causal relationship inferred by a model is not valid, calling into question decisions made based on the model. Prediction accuracy must be concordant with the realities of society, not just attempt to replicate predictions based on historically biased data. Our quest must be to find a better way to do this. To this, some researchers have proposed solutions to address Bias.

Reliability: Reliability indicates whether a model consistently generates the same results within the bounds of acceptable statistical error. Techniques designed to mitigate overfitting (e.g., regularization) and to adequately conduct model selection in the face of the bias/variance tradeoff can increase model reliability. Reliability measures may give insight into the risks related to decontextualization, especially when using dated or biased data leading to behaviors disconnected from the social contexts.

Robustness: Robustness is a measure of model sensitivity. Robustness contributes to sensitivity analysis in the risk management process. New robustness measures need to be developed as Machine learning, and AI algorithms learn from new data and user behavior. How the model changes as biases are introduced needs to be monitored. Data used by AI algorithms need to be monitored and audited to ensure that AI systems are robust.

Resilience and Security: AI systems must be resilient to adversarial AI attacks. There is a growing body of academic literature in this area, and I will refer the readers to that body of work.

Designers and develop control system characteristics. And systems should be built and tested, keeping the tenets of the Blueprint for the AI Bill of Rights [1]. Technical features include addressing the tradeoffs between convergent and

Table 2. Sample of existing literature

Paper	Types of Bias mentioned and Solutions offered to mitigate the Bias
Mehrabi, N., Morstatter, F., Saxena, N., Lerman, K., & Galstyan, A. (2021). A survey on Bias and fairness in machine learning. ACM Computing Surveys (CSUR), 54(6), 1-35.	Types of Bias: Measurement Bias, Omitted Variable Bias, Representation Bias, Aggregation Bias, Sampling Bias, Algorithmic Bias, User Interaction Bias, Historical Bias, Population Bias, Self-Selection Bias, Social Bias, Behavioral Bias, Temporal Bias, and Content Production Bias. Solutions: Designing Fair ML (Hard vs. soft debiasing). Fair Representation Learning (Variational Auto Encoders), Community Detection algorithms to understand customer base (Fair Machine Learning)
Mitchell, S., Potash, E., Barocas, S., D'Amour, A., & Lum, K. (2021). Algorithmic fairness: Choices, assumptions, and definitions. Annual Review of Statistics and Its Application, 8, 141-163.	Types of Bias: Statistical Bias, Societal Bias Solutions: Metric fairness is crucial when it comes to eliminating these kinds of biases, in other words choosing the right metric to evaluate the machine learning model. Finally, the machine learning model must be able to provide meaningful causal definitions.
Caton, S., & Haas, C. (2020). Fairness in machine learning: A survey. arXiv preprint arXiv:2010.04053.	Types of Bias: In processing, pre and post-processing modeling fairness issues at each stage of the build and deployment of machine learning models Solutions: Some of the key dilemmas the paper suggests to keep in mind (tradeoffs to know which Bias is ok to introduce). Dilemma 1: Fairness vs. Model Performance; Dilemma 2: (Dis)agreement and Incompatibility of "Fairness"; Dilemma 3: Tensions with Context and Policy; Dilemma 4: Democratization of ML vs. the Fairness Skills Gap
Jo, E. S., & Gebru, T. (2020, January). Lessons from archives: Strategies for collecting sociocultural data in machine learning. In Proceedings of the 2020 conference on fairness, accountability, and Transparency (pp. 306-316).	Types of Bias: Data consent, inclusivity, data power, data transparency, and ethics and privacy should be kept in mind while developing fair ML models. Solutions: Consent: (1) Institute data-gathering outreach programs to actively collect underrepresented data (2) Adopt crowdsourcing models that collect open-ended responses from participants and give them options to denote sensitivity and access Inclusivity: (1) Complement datasets with "Mission Statements" that signal commitment to stated concepts/topics/groups (2) "Open" data sets to promote ongoing collection following mission statements:- Power: (1) Form data consortia where data centers of various sizes can share resources and the cost burdens of data collection and management Transparency:(1) Keep process records of materials added to or selected from the dataset. (2) Adopt a multi-layer, multi-person data supervision system. Ethics & Privacy: (1) Promote data collection as a full-time professional career. (2) Form or integrate existing global/national organizations in instituting standardized codes of ethics/conduct and procedures to review violations

(continued)

Table 2. (*continued*)

Paper	Types of Bias mentioned and Solutions offered to mitigate the Bias
Cirillo, D., Catuara-Solarz, S., Morey, C., Guney, E., Subirats, L., Mellino, S., ... & Mavridis, N. (2020). Sex and gender differences and biases in artificial intelligence for biomedicine and healthcare. NPJ digital medicine, 3(1), 1-11.	Types of Bias: Healthcare bias Solutions: Reducing undesirable biases: (1) Stigma and discrimination (2) Unrepresentative samples (3) Adverse Reactions and Ineffective treatments Enhancing Desirable Biases: (1) Understanding Sexual and gender differences (2) Effective treatments for each sex/gender (3) Wellbeing of patients
Ehsan, U., Liao, Q. V., Muller, M., Riedl, M. O., & Weisz, J. D. (2021, May). Expanding explainability: Towards social Transparency in ai systems. In Proceedings of the 2021 CHI Conference on Human Factors in Computing Systems (pp. 1-19).	The paper discusses strategies for mitigating Sociotechnical Bias in ML models. Their work provides three levels of context and two kinds of effects. Their work also calls for the calibration of AI trust.
Shah, D., Schwartz, H. A., & Hovy, D. (2019). Predictive biases in natural language processing models: A conceptual framework and overview. arXiv preprint arXiv:1912.11078.	Types of bias: Eliminating Predictive Bias in ML models (Over amplification, label bias, semantic Bias, selection bias, error disparity, outcome disparity) Solutions: Understanding disagreements between labeling annotators can help improve labeling bias, re-stratifying the data to more closely match the ideal distribution can help mitigate selection bias, overamplification can be mitigated using synthetically matched distributions, de-bias embeddings can help with semantic Bias (hard debiasing, and soft debiasing), social level mitigation (spreadsheets to cover the traceability and Lifecycle of the dataset to understand origins and avoid historical Bias)

discriminant validity (whether the data reflects what the user intends to measure and no other things) and statistical reliability (whether the data may be subject to high levels of statistical noise and measurement bias). Explicit measures (automated, of course) based on variations of standard statistical techniques need to be in place. Further, data generated from experiments designed to evaluate system performance also fall into this category and might include tests of causal hypotheses and assessments of robustness to adversarial attack.

4.2   Socio-Technical Characteristics

Socio-Technical characteristics address (a) Explainability, (b) Interpretability, (c) Safety, (d) Privacy, and (e) Bias. We provide a brief on each of these characteristics in the following sub-sections

Explainability: For users to trust the system, it is essential to explain its functions. Explainability seeks to provide a programmatic, sometimes causal, description of how model predictions are generated. For example, if the coefficients cannot be explained in loan granting, it becomes difficult to explain how the different factors were considered in rejecting a loan application. This may lead some communities to wonder if the system is intrinsically unfair. Therefore, legislation and programming standard bodies must require programmers, system developers, and business managers to articulate clearly how the different factors used by the algorithm account for the outcomes.

Interpretability: Interpretability refers to the meaning of its output in the context of its designed functional purpose. Communicating the interpretation intended by model designers often alleviates risk. That said, understanding happens in the social context without proper communication, considering beliefs individuals and communities hold. Therefore, it is essential to communicate at a level concordant with the level at which a regular user understands the message.

Privacy: Privacy generally refers to the norms and practices that help to safeguard values such as human autonomy and dignity. Disclosures must always happen with consent. If they have not already done so, many countries are developing privacy legislation to afford their citizens rights in the Wild West of the internet and automated systems.

Safety: Safety as a concept is highly correlated with risk and generally denotes an absence (or minimization) of failures or conditions that render a system dangerous. For example, applications developed in the healthcare space should be tested regarding patient safety before use. Furthermore, they should embody the principle of "Do no harm." Therefore, societies must create bodies to test Software to prevent harm.

Bias: While Bias is not always a negative phenomenon, certain biases exhibited in AI models and systems can perpetuate and amplify negative impacts on individuals, organizations, and society, and at speed and scale far beyond the traditional discriminatory practices that can result from implicit human or systemic biases.

5   Conclusion

5.1   Need for Research

To ensure a future free from tyranny from badly created AI applications, we must think of the issues elaborated on in this article. We are all in the business of developing, measuring and using AI systems. Our realizations must translate to how we build AI that enhances the human experience. This requires us to embody the principle 'DO NO HARM' - a code that physicians call the 'Hippocratic Oath.'

From a research perspective, a growing body of literature focuses on bias identification and the type of Bias. However, no design theory literature or Design Science Research addresses the issue end-to-end and how to build, validate, deploy and manage AI systems taking into account the framework by NIST.

Much research needs to be done informing the community (a) how to develop data that forms the input to AI systems, (b) How to build AI systems free from

Bias, (c) how to validate the system, (d) how to correct the systems and (e) how to manage systems on an on-going basis. There is some re- search addressing Machine Learning bias. Table 2 provides a very brief account of that literature (note: this table is not a literature survey, nor is it an exhaustive list, it is, however, a sample of the literature). Research work in all the areas concerning Bias in AI needs to be better informed by future research. The hope is that prospective doctoral students will take up these issues in a very domain-specific manner. From a management perspective, research is needed on the cost and pricing of building safe systems and on audit and governance matters. And how will these costs impact the consumer? What new businesses will this quest for better AI generate?

5.2   Need for Legislation

We do not allow cars that are not roadworthy to be driven on our roads. We do not allow people without the proper license to operate. We must do the same with Automated Systems. Most governments recognize the privacy issue and are creating laws to safeguard the public. We need to adopt the same practice for AI and automated systems so they do not operate in destructive ways. This is extremely important not only in the Healthcare area but in all areas. That said, we have a good beginning in the Healthcare area in the United States. The USA Food and Drug Administration now processes Healthcare applications and certifies them. Yes, these additional processes cause delays and impact costs. But your life is worth it. Table 3, adapted from the (IMDRF Working Group, 2014) [6] document on Software as a Medical Device (SaMD), provides some guidance on risk characterization. The link to that website is included as part of the Table legend. This table allows developers to classify their healthcare applications based on risk levels. Sure more always need to be done. Software developers must select and implement adequate processes for the planning, design, development, deployment, and documentation of robust and dependable software in the healthcare arena commensurate with risk. This regulation is a good beginning. We need to do more in terms of guidance.

That said, how we train developers has to change. New textbooks are needed to educate the future workforce on bias and AI issues.

Table 3. FDA risk categories for SaM [10]

State of Healthcare situation or condition	Significance of Information provided by SaMD		
	Treat of Diagnose	Drive Clinical Management	Inform Clinical Management
Crucial	High Impact	High Impact	Medium Impact
Serious	High Impact	Medium Impact	Low Impact
Non-Serious	Medium Impact	Low Impact	Low Impact

Acknowledgments. First, thanks go to two colleagues Dr. Cristian Tiu and Dean Ananth Iyer from the University of Buffalo for discussing the topic of Bias in AI. Their conversations have made me an advocate for change. I also want to thank two of my doctoral students, Ms. Sagarika Suresh Thimmanayanapalya and Mr. Raghvendra Singh, for their contributions to developing this manuscript. Finally, I must thank the following MS in MIS students at the University at Buffalo: Vedant Srivastava, Kajol Govind Vaghela, Harshada Dayasagar Samant, and Teja Sai Ram Jonadula for their help.

References

1. AI bill of rights. https://www.whitehouse.gov/ostp/ai-bill-of-rights/. Accessed 19 Nov 2022
2. AI market research. https://www.grandviewresearch.com/industry-analysis/artificial-intelligence-ai-market. Accessed 19 Nov 2022
3. Big data analytics market size. https://www.fortunebusinessinsights.com/toc/big-data-analytics-market-106179. Accessed 19 Nov 2022
4. Coordinated plan on artificial intelligence 2021 review. https://digital-strategy.ec.europa.eu/en/library/coordinated-plan-artificial-intelligence-2021-review. Accessed 19 Nov 2022
5. European approach to AI. https://digital-strategy.ec.europa.eu/en/policies/european-approach-artificial-intelligence. Accessed 19 Nov 2022
6. IMDRF/SaMD WG/N12FINAL (2014). https://www.imdrf.org/sites/default/files/docs/imdrf/final/technical/imdrf-tech-140918-samd-framework-risk-categorization-141013.pdf. Accessed Nov 2022
7. Managing the risks of artificial intelligence. https://www.nist.gov/speech-testimony/trustworthy-ai-managing-risks-artificial-intelligence. Accessed 19 Nov 2022
8. Nist. https://www.nist.gov/itl/ai-risk-management-framework. Accessed 19 Nov 2022
9. Nist special publication. https://nvlpubs.nist.gov/nistpubs/SpecialPublications/NIST.SP.1270.pdf. Accessed 19 Nov 2022
10. "software as a medical device": Possible framework for risk categorization and corresponding considerations. https://www.imdrf.org/sites/default/files/docs/imdrf/final/technical/imdrf-tech-140918-samd-framework-risk-categorization-141013.pdf. Accessed 19 Nov 2022
11. Types of statistical bias. https://online.hbs.edu/blog/post/types-of-statistical-bias. Accessed 19 Nov 2022
12. Arfè, A., et al.: Non-steroidal anti-inflammatory drugs and risk of heart failure in four European countries: nested case-control study. BMJ **354** (2016)
13. Elston, D.M.: Survivorship bias. J. Am. Acad. Dermatol. (2021)
14. Faillie, J.L.: Indication bias or protopathic bias? Br. J. Clin. Pharmacol. **80**(4), 779–780 (2015)
15. Jargowsky, P.: Omitted Variable Bias. Encycl. Soc. Meas. **2**, 919–924 (2005)
16. Tripepi, G., Jager, K.J., Dekker, F.W., Zoccali, C.: Selection bias and information bias in clinical research. Nephron Clin. Pract. **115**(2), c94–c99 (2010)

Big Data in Cognitive Neuroscience: Opportunities and Challenges

Kamalaker Dadi[1] and Bapi Raju Surampudi[1,2(✉)]

[1] IHub-Data, International Institute of Information Technology,
Hyderabad 500032, India
kamalaker.dadi@ihub-data.iiit.ac.in
[2] Cognitive Science lab, International Institute of Information Technology,
Hyderabad 500032, India
raju.bapi@iiit.ac.in

Abstract. Cognitive brain mapping is enjoying its growth with the availability of large open data sharing efforts as well as the application of modern machine learning and deep learning methods. In this article, we review some of the current practices in cognitive neuroscience predominantly focusing on functional imaging and highlight the tremendous opportunities fostered by the unprecedented scale of datasets in cognitive neuroscience. We also point out challenges and limitations to keep in mind while working with these datasets.

Keywords: Cognitive science · Brain function · Neuroimaging · Encoding · Machine learning · Decoding

1 Introduction

Cognitive neuroscience is an interdisciplinary field that studies the neural bases of mental abilities. How does the neuroanatomical structure in the brain give rise to mental functions, such as perception, learning and memory, vision, language, and reasoning? What are the neural mechanisms that are engaged in a memory task? What are the neural mechanisms that play a differential role in monolingual or bilingual brains in language comprehension or production? These are a few questions that can be addressed by designing well-controlled experiments that recruit specialized brain regions to specific mental functions (functional specialization) [1]. There has been an exponential growth in the number of studies over the past couple of decades using modern neuroimaging methods along with various inferential strategies and conceptual frameworks to infer brain-behavior relationships, resulting in one brain region mapped to one or more mental functions [2]. For example, hippocampus is observed to be a complex brain region that has been mapped to many mental functions such as episodic memory, declarative memory, recollection, novelty detection, spatial navigation, and even more complex functions leading to formation of cognitive maps. Apart from such focused investigations, in the last decade (as part of the *decade of the brain* theme),

© The Author(s), under exclusive license to Springer Nature Switzerland AG 2022
P. P. Roy et al. (Eds.): BDA 2022 India, LNCS 13773, pp. 16–30, 2022.
https://doi.org/10.1007/978-3-031-24094-2_2

several task-free large-scale neuroimaging datasets have been created (see for example, Table 1. With the advent of large-scale datasets in cognitive neuro-science and combining the insights from focused studies, the grand challenge is to see if we can start building the cognitive functional architecture of the brain. In this paper, we review the current practices in cognitive neuroscience, existing datasets and some opportunities and challenges in cognitive neuroscience.

1.1 Functional Segregation and Functional Integration

The localization of a brain structure/region to a particular mental function was first formulated by Gall in the early 19th century, later referred to as Gall "phrenology". Since then, the identification of a brain region to a specific func-tion has become a central theme in Neuroscience. It was in 1881, based on observations from patients with language difficulty, Broca and Wernicke identi-fied neuroanatomical locations of brain damage (disconnections) and map them to specific behavioral impairments, for e.g., language comprehension and pro-duction. Further investigations revealed how anatomical structures might be working together to support mental functions, apart from individual functions being related to specific brain regions. The current view is that brain function is mediated by two fundamental principles: segregation (anatomically segregated) and integration (functionally dependent or connectivity) [1]. Segregation refers to sub-division of the whole brain to specialized units/modules e.g., cortical and sub-cortical structures segregated among themselves based upon their neuronal properties. On the other hand, integration refers to multiple brain regions and the resulting dynamic interaction among different mental capacities that give rise to a particular cognitive, sensory, or motor function. From a methodologi-cal perspective, understanding such structure-function relations is still an active research topic in neuroscience, e.g., investigating using graph theoretic meth-ods [3] and its relation to cognition [4].

2 Inferential Approaches in Cognitive Neuroscience

Traditionally, assigning functions to brain regions has been based on approaches from many different disciplines that are interested in the study of the mind and behavior. We briefly review some of them.

Lesion-Behavior Mapping

Lesion-behavior mapping predicts the deficit in the mental function given the lesion in the brain region. For example, with this approach, Brenda Milner and colleagues observed a crucial role played by hippocampus in the foundation of new memories following a damage to the medial temporal lobe - an area where hippocampus resides in the brain [5]. Thus, this inferential approach identifies behavioral function given focal damage to the brain. Contrary to neuroimaging methods that look at whole brain activation, one of its strengths is that it allows to infer causal relationships between brain and behavior [6]. Moreover, this app-roach is clinically useful due to its ability to predict performance in modeling post-stroke outcomes-based lesion imaging. However, the fact that the brain is functionally organized into larger dynamic networks may place severe limitations to the inferential power of this approach [2].

Imaging-Based Approaches

Neuroimaging techniques such as electro-encephalography (EEG), positron emission tomography (PET), functional Magnetic Resonance Imaging (fMRI), and magneto encephalography (MEG) have become crucial investigation tools for the field of cognitive neuroscience. They reveal the brain regional activations in response to mental activities and most of these are non-invasive. The first study that took the advantage of PET was in late 1980s that used a block design that alternated between one block (condition of interest) and another block (control). Using simple subtraction techniques between the two conditions, one can begin to localize cognitive functions in the brain [1]. MRI based upon the blood-oxygen-level-dependent (BOLD) technique [7] also started with block design experiments and by end of the 1990s. fMRI (non-invasive) quickly became popular over PET (invasive), especially with the addition of event-related experimental designs that allow investigation of evoked responses similar to electrophysiology [8].

With fMRI, even simpler tasks like finger tapping give a detailed characterization of the activation over widespread areas of the brain. From there, new designs and better conceptualizations strive to address the core principles of brain functional organization, in particular functionally distributed networks [9]. It is possible to investigate how distributed brain regions co-vary while performing a task, as characterized in terms of functional connectivity and effective connectivity (the causal influence one region on another) [10]. Moreover, fMRI is also useful for clinical cases, for instance, studying the reorganization of brain plasticity in response to disease and other abnormalities.

The Stimulation Approach

Stimulation techniques such as transcranial magnetic stimulation (TMS), transcranial direct current stimulation (tDCS) work in a similar way like lesion-behavior mapping approaches, where temporary (reversible) lesions are created by these modern stimulation techniques. TMS uses very brief high intensity magnetic fields to induce currents and thus depolarize neurons to either alter (lesion like impairments) or increase (cognitive performance) in regions of the brain [11]. After brief brain stimulation, the cause of any change in behavior is examined. The duration of the TMS pulse for excitation, uncontrolled spread of stimulation effects to spatially distant regions, and its ability to study only superficial brain regions may hinder the use of this approach for cognitive neuroscience [2]. In addition, studies have begun to explore the usefulness of these stimulation techniques for rehabilitation of brain function by inducing local plasticity [11].

The Structure-Behavior Approach

This approach correlates the morphometric features of the brain structures like the local gray matter volume and cortical thickness, other tissue types to neuropathologies and behavior. For example, local gray matter volume and its variations are studied in relation to brain ageing or general intelligence ('g') [12].

3   Current Practices in Cognitive Neuroscience

Mapping the function of each brain region entails linking brain systems to a wide variety of cognitive tasks [13]. fMRI has proven to be a dominant tool in cognitive neuroscience[1]. Current practices in data aggregation and knowledge accumulation bring in novel opportunities for a systematic characterization of brain regions across a range of behavioral conditions and phenotypic features. We first review such practices with mapping tools like – task-fMRI, meta-analyses, mega-analyses [14], intra-individual analysis [15], population imaging with resting-state fMRI [16] – where each tool plays a pivotal role in cognitive neuroscience.

Task-fMRI is used to map BOLD signal changes in brain areas underlying the performance of certain (cognitive) tasks. Activation maps (effects of interest) are then obtained by mapping task manipulations to brain activity separately for each voxel or region with an encoding analysis. It models the BOLD signal as a linear combination of experimental conditions—the General Linear Model (GLM, [18]). The BOLD signal forms a matrix $Y \in \mathbb{R}^{n \times p}$, where p is the number of voxels, n is the number of timepoints/samples. The GLM models Y as

$$Y = X\beta + \epsilon \tag{1}$$

where $X \in \mathbb{R}^{n \times q}$ is the design matrix formed by q temporal regressors which denote conditions of interest (e.g., visual stimuli versus rest) or nuisance variables and ϵ is noise [1]. Typically, the estimated activation maps ($\hat{\beta}$-maps) elicited by the task (psychological) conditions should provide a mapping of the targeted mental process. Finally, task representative statistical maps are either interpreted in terms of peaks of activation or shared as functional contrasts in public repositories, like NeuroVault [19]. Figure 1 shows the schematic overview of estimating such activation maps using encoding framework on task-fMRI images.

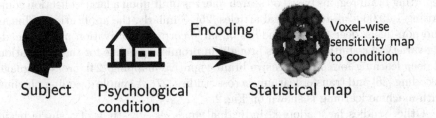

Fig. 1. Schematic of task-fMRI analysis. In this example, during fMRI scanning the subject is viewing the condition of interest, i.e., house stimuli and the resulting regional activations are mapped onto the brain showing activation in visual areas using standard encoding frameworks.

Table 1 lists several task-fMRI datasets, their objectives, cognitive batteries, the number of subjects, experimental tasks, and other demographic details. As

---

[1] In 2010 alone, over 1,000 fMRI articles were published [17].

can be seen from the table the breadth of cognitive experiments covered by each study is small to moderate to facilitate thorough demarcation of brain regions. However, a sizable collection of them would help in piecing together the overall functional architecture of the brain.

Data accumulation is a long-standing and ongoing effort in cognitive neuroimaging that allows us to map brain locations to mental operations at large-scale [20]. This can be achieved by combining responses to multiple tasks by pooling in activation results from many task-fMRI studies. On top of fine demarcation of brain regions, these pooling efforts also bring in several additional benefits. Such as detecting false positives as they appear by chance hence do not replicate across pooled data and also increase in the brain mapping accuracy [21].

With the increase in the brain mapping literature, there is also an increase in the data sharing efforts towards reproducible science in various disciplines. Thanks to those databases that store petabytes of data coming in various formats such as: raw 4D timeseries data[2], cognitive ontologies[3] [22], processed or analyzed data indicating peak activation coordinates or activation maps coming from neuroimaging literature and sharing them via BrainMap[4] [23] or in repositories like Neurovault[5] [19] or Neurosynth[6] which currently account for a whopping 14,371 studies in its database [17]. These initiatives open up the chance of developing robust analytic methods. Having access to shared information, analytic methods such as meta- (pooled analysis on individual publications) or mega-analyses (pooled analysis on images) could map the brain locations that showed significant consistency with the links to mental functions [24].

Recent years have seen the application of machine learning and deep learning methods for meta- & mega-analyses that utilize databases like Neurosynth and Neurovault at large-scale. It is possible to fire a query about possible activated brain regions and other quantities using Neuroquery[7]. For example, a query with a term like *emotion* predicted amygdala (see Table 2). This dominant style of reporting brain regions based on search query is built upon a large collection comprising 1,49,000 full-text journal articles [25]. Similarly, the application of linear and non-linear deep learning models on large corpus of activation maps, decoding variety of cognitive concepts brought in promising avenue for the application of deep learning for comprehensive brain cognitive mapping [24], cross-language decoding [26] and transfer learning across studies [27]. A simple decoding pipeline with machine learning is shown on Fig. 2.

Other studies have adopted individual analyses – a study of a single brain in detail or deep phenotyping of individuals across more than 29 tasks (ongoing project on 12 participants) that cover wide range of cognitive systems for indi-

---

[2] https://openneuro.org/.
[3] https://www.cognitiveatlas.org/.
[4] https://www.brainmap.org/.
[5] https://neurovault.org/.
[6] https://neurosynth.org/.
[7] https://neuroquery.org/.

Table 1. List of few openly accessible task-fMRI datasets and the cognitive neuroscience experiments covered by each dataset.

Project	Task-fMRI datasets			
	Objective	Subjects	#Tasks	Cog. Domain
HCP	Study patterns of structural and functional connectivity in the healthy subjects and its alterations in cognitive and behavioral variables	1200 (22–35 years)	Retinotopic Mapping Finger Responses	Visual, Somatosensory Motor
			N-back Task (2-back versus 0-back)	Working Memory; Cognitive Control
			Auditory sentence and story	Language Processing
			Negative & neutral faces	Emotion Processing
			Relational & matching features	Relational Processing
			Gambling task Reward & Loss	Reward & Decision making
			Social & random interactions	Social & cognition
Cam-CAN	To understand how age-related changes to neural structure and function interact to support cognitive abilities across the lifespan	700 (18–87 years)	Audio visual stimuli & manual response	Sensorimotor
			Watch & listen to movie Alfred Hitchcock's "Bang! You're Dead"	Stimulus driven mechanisms
The UK Biobank	Population study	40,000 (40–69)	Matching shapes & emotional faces	Emotion Processing

HCP – Human Connectome Project (https://www.humanconnectome.org/study/hcp-young-adult), Cam-CAN – Cambridge Center for Ageing in Neuroscience (https://www.cam-can.org/index.php?content=dataset), UK Biobank – United Kingdom Biobank (https://www.ukbiobank.ac.uk/enable-your-research/about-our-data/imaging-data)

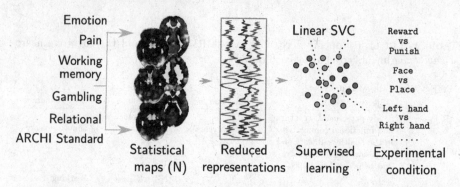

Fig. 2. Inter-subject decoding pipeline. A machine learning algorithm can leverage activation maps from multiple studies and perform inter-subject decoding across different experimental conditions. Inputs are the activation maps from many task-fMRI studies that are then projected onto dictionaries of functional networks e.g., ICA networks. Then, these reduced representations are passed to linear support vector classifier (SVC) for classification. The learned model could now be used to decode brain activation map of a test subject to decipher the experimental condition (class label) associated with the brain activation.

Table 2. A simple query like "emotion" to Neuroquery.org predicted amygdala region with more weight than closely-related other terms

Term	Weights assigned to brain regions	No. of studies containing the term
amygdala	1.00	4540
face	0.18	4405
insula	0.14	7050
prefrontal	0.11	10017
fusiform gyrus	0.04	3445
hippocampus	0.04	4610

vidual functional atlasing [28]. The advantage of such individual-specific projects is that they begin to establish unequivocal relationship between functional specialization of brain regions, providing rich and heterogeneous patterns of region-behavior associations at individual level and therefore significantly contribute to understanding the brain-bases of human cognition. Another example dataset in the vein of intra-individual task-fMRI is *studyforrest* data which is multimodal and the task here is to listen to the audio descriptions from the Forest Gump movie. This dataset employs complex naturalistic stimuli to encode complex auditory and visual information like the ability to perceive language, music and other visual properties of the stimuli [29]. A list of datasets that support neuro-cognitive modeling are shown in Tables 3 and 4, respectively.

Table 3. List of few openly accessible task-fMRI datasets for intra-subject cognitive mapping. For more details about the experimental protocols in IBC can be seen in their respective papers.

Project	Task-fMRI datasets for intra-individual analysis			
	Subjects	Battery	#Tasks	Cog. Domain
IBC	12 (26–40 years)	ARCHI	Standard Spatial Social Emotion	Visuomotor language arithmetic social emotional
		HCP	Emotion Gambling Motor Language Relational Social Working Memory	Motor emotional social relational gambling working memory
		RSVP	Language	Sentence comprehension
		MTT	West-East South-North	Space/ Time representation
		Preference	food items paintings human faces and houses	Pleasantness Rating Incentive Reward
		TOM	TOM Localizer TOM and Pain-Matrix Narrative Localizer TOM and Pain-Matrix Movie Localizer	Representation of beliefs, facts, observed pain
		VSTM	vstm response to constant numerosity	Short-term memory, numerosity
		Enumeration	enumeration response to constant numerosity	Enumeration, numerosity
		Self	self-reference encoding other-reference encoding	Encoding and retrieving representation of self and others
		Bang	speech no-speech	Unconstrained audio-visual stimulation
		Retinotopy	Wedge Ring	Retinotopy

Individual Brain Charting (IBC) project link for more details https://project.inria.fr/IBC/files/2020/10/documentation_vs_3-2.pdf; RSVP – Rapid Serial Visual Presentation; MTT – Mental Time Travel; TOM – Theory of Mind; VSTM – Visual Short-Term Memory

Table 4. Naturalistic stimuli and dense sampling fMRI datasets

Dataset	#Subjects	Type of study	Cog. Domain
Narratives [a]	345 (18–53) years old	Listen to 27 diverse stories	language and narrative comprehension
StudyForrest[b]	20 (21–38) years old	Auditory feature film (Forrest Gump)	auditory attention and cognition, language and music perception, and social perception
Courtois Neuromod[c] 100+ hours of functional neuroimaging per subject	6 (31–47) years old	HCP test-retest each subject repeated for 15 times	restingstate, gambling, motor, social, working memory, emotion, language and relational
		movie10 Bourne Supremacy, The Wolf of Wall Street, Life documentary (twice), Hidden Figures (twice) Friends TV show season 1 & 2	Disorders Traits Behavior
The Natural Scenes Dataset (NSD)[d]	8 (19–32) years old	View 9,000–10,000 color natural scene	low-level high-level vision memory connectivity
BOLD5000[e]	4	4916 unique scene images	Visual scene processing
Midnight Scan Club[f]	10 (24–34) years old	face monitor/discrimination abstract/concrete task dot motion task semantic decision task motor task paradigm rest eyes open	Motor Perceptual & Language Incidental memory

[a] https://openneuro.org/datasets/ds002345
[b] http://studyforrest.org
[c] https://docs.cneuromod.ca/_/downloads/en/latest/pdf/
[d] http://naturalscenesdataset.org/
[e] https://bold5000-dataset.github.io/website/
[f] https://openneuro.org/datasets/ds000224/

On the other hand, a further line of research relies on resting-state fMRI (rsfMRI) data with the purpose of characterizing spontaneous, intrinsic activation of brain regions and networks, namely, connectomes [30]. There is a tremendous increase in the development of population imaging data cohorts for rsfMRI [31]. They provide multimodal brain imaging data, current lifestyle, questionnaire scores of personality and other neuropsychological measures that cover several cognitive domains such as emotion, memory and learning. Such data cohorts have been used for variety of reasons such as developing methods for functional connectomes [32], studying various aspects of the brain-behavior relationships [33], and brain diseases [34]. One of the current challenges is characterizing the relation between resting-state functional connectivity and the underlying structural connectivity [3].

4   Opportunities

In the previous section, we have hinted at the applications of machine learning in leveraging such vast volumes of data to gain insight into brain cognition. Machine learning techniques greatly facilitate the analysis on large cohorts by making the data computationally tractable (low-dimensional representations), mitigate human biases, easier to share across laboratories, and requiring less disk storage. For example, in the decoding pipeline shown in Fig. 2, the framework is largely benefited from machine learning approaches to make the analysis computationally tractable on large-scale activation maps aggregated across studies. After aggregation, the next step is to reduce the high-dimensional brain images to low-dimensional representations that summarize the information present in the images while keeping meaningful representations of the brain *a.k.a* data reduction. Brain atlases (originated from Brodmann's labeling of areas) are widely used for data reduction in functional imaging. Brain atlases can be reference anatomical atlases such as Automated Anatomical Labeling (AAL) [35] or atlases estimated using data-driven approaches based on k-means or Ward clustering, as well as Independent Component Analysis (ICA) or dictionary learning [36]. There are great benefits in learning brain regions from the neuroimaging data than pre-defined anatomical atlases, as the choice of brain regions defines the brain representations for further analysis in the framework [37]. Analyzing structural MRI can also be seen as application domain for machine learning where automated segmentation of labeled regions from structural MRI scans [38] and white matter tracts from diffusion MRI [39] are useful.

Another potential application of machine learning technique is the automated annotation of raw EEG signals for sleep-staging [40]. In this example, self-supervised learning is used to aggregate complementary information available from multiple inputs for unsupervised learning of EEG representations. Such annotations are usually accomplished manually, the development of these techniques immensely reduces the involvement of human time and effort, mitigate human biases or error in annotating signals. By training these techniques on large corpus of collections will make the results more stable and reproducible. These techniques have the scope beyond signal processing, for example for annotating behavior from the video recordings.

Machine learning not only allows us to extract structured information, but also enables learning data-driven taxonomy. A fMRI study by Chang et al. [41] identified neural signatures sensitive and specific to negative emotion and pain with machine learning deriving the data-driven taxonomies of affective processes. Beyond decoding states, machine learning can yield testable hypothesis. For example, if two brain regions are identified to be the predictors of stress induced behavior, how are those two regions functionally connected? These questions can further be confirmed via additional experiments that have some capabilities of identifying the causal mechanisms of action potential patterns in these neurons of two specific regions, for example using optogenetic experiments [42].

The correspondence between the working principles of the brain and their resemblance to the principles of machine learning is an active area of research.

Just as visual neurons are sensitive to orientation of visual gratings in the primary visual cortex (V1), early layers of the deep neural networks seem to learn the filters that capture this simple orientation. Whereas response in nodes in deeper layers show resemblance to higher-order features of natural objects and categories observed in the primate ventral stream areas (V1-V3) as well as Inferior temporal cortex (IT). Deep learning has now made enormous developments and advancements in the statistical and computational frameworks such that they can yield new theories of brain function [43]. Currently data generation efforts are leading the theory-making efforts. Thus future theories of brain function can be appropriately constrained by large datasets to test themselves on. Theory-making is important so that we can simulate and test how the function breaks down and leads to neurodegenerative or pyschiatric disorders [44]. The other opportunity is to utilize big datasets related to Movie Watching, Passive Reading/Listening (see Tables 1, 3, and 4) and investigating how these relate to controlled studies of visual/auditory/linguistic functions? Eventually it should be possible to unravel brain networks and dynamics when one is experiencing a narrative (movie, story reading or listening) and use such large datasets to do computational experiments at scale that are not feasible empirically [45].

5   Challenges

This section highlights some challenges and current limitations in effectively utilizing big datasets in cognitive neuroscience.

Noise in the fMRI Signal

fMRI raw timeseries signal is not by itself separated from noise sources, whether physiological or acquisition-related [46]. Moreover, subjects may feel uncomfortable while undergoing scanning and often move while sampling BOLD time series. This motion-related time series may create spurious functional connectivity or may lead to false positives in group studies if not addressed properly [47]. These effects are a typical confound when linking brain activity to pychological conditions. Another well known issue is subjects falling asleep during acquisition due to boredom which strongly corrupts the recorded signal [48].

Several denoising methods and pipelines have already been proposed for cleaning such noisy BOLD timeseries signals [49]. These methods can extract nuisance signals on whole-brain voxels, capturing physiological noise and subject-motion signals using common data pre-processing procedures on raw fMRI data. The estimated noise signals can be combined together as nuisance covariates in GLM approach (Eq. 1) to regress out nuisance signals when estimating brain activation maps.

Sources of Variability

Another major challenge is the inter-subject variability [50], for example, subject-to-subject variations in the spatial locations of task-evoked responses. When the functional brain topography of each subject is compared to group-averaged topography, they show a consistent inter-subject variability. This comes from the fact that individuals might activate distinct brain networks and topology in response to the same stimuli due to differences in the brain shape, size, life experiences, context, etc. [51]. To overcome this challenge, efforts are in place for dense sampling of fMRI data with a large set of cognitive tasks on the same cluster of individuals for precision functional mapping [28].

From a group-analysis standpoint, different brains (across subjects) are warped into a common space (i.e., registering subject space to common space via MNI152 template), cortical folding patterns based registration, such that it yields meaningful functional comparisons. Recent availability of task-free resting-state data as well as natural tasks like movie-watching, with the inclusion of variety of rich stimulus conditions and identification of fMRI patterns that are correlative across subjects (Inter-subject correlation) opens new avenues for bringing subjects into a common space to facilitate alignment of function across subjects [52]. Other methods include, multimodal surface matching framework which is introduced to match the combination of any multi-modal features [53], hyper-alignment that maps the patterns of activation in each participant into a common representational space such that the group of voxels that share common functional properties share the same location in the high-dimensional space and similarly, the representational dissimilarity matrices and shared response modeling [52]. From an inter-individual differences standpoint, such topographies are also important for tracking health and identifying disease [16].

Test-Retest Reliability

The reliability of fMRI data depends on various factors such as the task subjects are doing, experimental design, analysis procedures, and length of acquisition of fMRI scans. Intra-class correlation coefficients (ICC) is a method very similar to Pearson's correlation, which measures test-retest reliability on repeated measurements (e.g., brain activity) of the same subject under the same conditions or task. ICC ranges between 0 and 1 with higher values indicating better reliability of subject's measurements.

6   Conclusion

With the advent of neuroimaging investigation of brain function in the last couple of decades, there are unprecedented opportunities of mapping the cognitive architecture of the brain. Large neuroimaging datasets are being made available publicly that span the whole spectrum of spontaneous brain activity during resting state, task-based functional activity, task-free experience of naturalistic stimuli, as well as brain structure and function from patients suffering from specific neurodegenerative or psychiatric conditions. A categorized list of such datasets is presented here. Various tools that enable systematic queries on the neuroscience databases are also briefly mentioned. Finally, the article outlines various opportunities as well as challenges of using such datasets.

Acknowledgments. KD acknowledges IHub-Data, IIIT Hyderabad for financial support.

References

1. Friston, K.J.: Imaging neuroscience: principles or maps? Proc. Nat. Acad. Sci. **95**(3), 796–802 (1998)
2. Genon, S., Reid, A., Langner, R., Amunts, K., Eickhoff, S.B.: How to characterize the function of a brain region. Trends Cogn. Sci. **22**(4), 350–364 (2018)
3. Surampudi, S.G., Misra, J., Deco, G., Bapi, R.S., Sharma, A., Roy, D.: Resting state dynamics meets anatomical structure: temporal multiple kernel learning (tMKL) model. Neuroimage **184**, 609–620 (2019)
4. Park, H.-J., Friston, K.: Structural and functional brain networks: from connections to cognition. Science **342**(6158), 1238411 (2013)
5. Scoville, W.B., Milner, B.: Loss of recent memory after bilateral hippocampal lesions. J. Neurol. Neurosurg. Psychiatry **20**(1), 11–21 (1957)
6. Rorden, C., Karnath, H.O.: Using human brain lesions to infer function: a relic from a past era in the fMRI age? Nat. Rev. Neurosci. **5**(10), 813–819 (2004)
7. Ogawa, S., Lee, T.M., Kay, A.R., Tank, D.W.: Brain magnetic resonance imaging with contrast dependent on blood oxygenation. Proc. Nat. Acad. Sci. **87**(24), 9868–9872 (1990)
8. Dolan, R.J.: Neuroimaging of cognition: past, present, and future. Neuron **60**(3), 496–502 (2008)
9. Friston, K.: Beyond phrenology: what can neuroimaging tell us about distributed circuitry? Ann. Rev. Neurosci. **25**(1), 221–250 (2002)
10. Friston, K.J.: Functional and effective connectivity: a review. Brain Connectivity **1**(1), 13–36 (2011)
11. Luber, B., Lisanby, S.H.: Enhancement of human cognitive performance using transcranial magnetic stimulation (TMS). Neuroimage **85**, 961–970 (2014)
12. Lerch, J.P., et al.: Studying neuroanatomy using MRI. Nat. Neurosci. **20**(3), 314–326 (2017)
13. Poldrack, R.A.: Mapping mental function to brain structure: how can cognitive neuroimaging succeed? Perspect. Psychol. Sci. **5**(6), 753–761 (2010)
14. Costafreda, S.: Pooling FMRI data: meta-analysis, mega-analysis and multi-center studies. Front. Neuroinformatics **3** (2009)
15. Nieto-Castañón, A., Fedorenko, E.: Subject-specific functional localizers increase sensitivity and functional resolution of multi-subject analyses. Neuroimage **63**(3), 1646–1669 (2012)
16. Biswal, B.B., Mennes, M., Zuo, X.-N., Gohel, S., Kelly, C., Smith, S.M., et al.: Toward discovery science of human brain function. Proc. Nat. Acad. Sci. **107**(10), 4734–4739 (2010)
17. Yarkoni, T., Poldrack, R.A., Nichols, T.E., Van Essen, D.C., Wager, T.D.: Large-scale automated synthesis of human functional neuroimaging data. Nat. Methods **8**, 665–670 (2011)
18. Friston, K.J., Holmes, A.P., Worsley, K.J., Poline, J.-B., Frith, C., Frackowiak, R.: Statistical parametric maps in functional imaging: a general linear approach. Hum. Brain Mapp. **2**(4), 189–210 (1995)
19. Gorgolewski, K.J., et al.: Neurovault.org: a web-based repository for collecting and sharing unthresholded statistical maps of the human brain. Front. Neuroinformatics **9**, 8 (2015)
20. Fox, P.T., Parsons, L.M., Lancaster, J.L.: Beyond the single study: function/location metanalysis in cognitive neuroimaging. Curr. Opin. Neurobiol. **8**(2), 178–187 (1998)

21. Gurevitch, J., Koricheva, J., Nakagawa, S., Stewart, G.: Meta-analysis and the science of research synthesis. Nature **555**(7695), 175–182 (2018)
22. Poldrack, R., et al.: The cognitive atlas: toward a knowledge foundation for cognitive neuroscience. Front. Neuroinformatics **5** (2011)
23. Fox, P.T., Lancaster, J.L.: Opinion: mapping context and content: the brainmap model. Nat. Rev. Neurosci. **3**, 319–321 (2002)
24. Wager, T.D., Lindquist, M., Kaplan, L.: Meta-analysis of functional neuroimaging data: current and future directions. Soc. Cognit. Affect. Neurosci. **2**(2), 150–158 (2007)
25. Dockès, J., et al.: Neuroquery, comprehensive meta-analysis of human brain mapping. eLife **9**, 53385 (2020)
26. Oota, S.R., Arora, J., Gupta, M., Bapi, R.S.: Multi-view and cross-view brain decoding. In: Proceedings of the 29th International Conference on Computational Linguistics (COLING), pp. 105–115 (2022)
27. Mensch, A., Mairal, J., Thirion, B., Varoquaux, G.: Extracting representations of cognition across neuroimaging studies improves brain decoding. PLoS Comput. Biol. **17**(5), 1–20 (2021)
28. Pinho, A.L., et al.: Subject-specific segregation of functional territories based on deep phenotyping. Hum. Brain Mapp. **42**, 841–870 (2021)
29. Sengupta, A., et al.: A studyforrest extension, retinotopic mapping and localization of higher visual areas. Sci. Data **3**, 160093 (2016)
30. Sporns, O., Tononi, G., Kotter, R.: The human connectome: a structural description of the human brain. PLoS Comput. Biol. **1**, 42 (2005)
31. Madan, C.R.: Scan once, analyse many: using large open-access neuroimaging datasets to understand the brain. Neuroinformatics **20**(1), 109–137 (2022)
32. Yeo, B.T.T., Krienen, F.M., Sepulcre, J., Sabuncu, M.R., et al.: The organization of the human cerebral cortex estimated by intrinsic functional connectivity. J. Neurophysiol. **106**, 1125 (2011)
33. Smith, S.M., et al.: A positive-negative mode of population covariation links brain connectivity, demographics and behavior. Nat. Neurosci. **18**(11), 1565–1567 (2015)
34. Greicius, M.: Resting-state functional connectivity in neuropsychiatric disorders. Curr. Opin. Neurol. **21**, 424 (2008)
35. Tzourio-Mazoyer, N., et al.: Automated anatomical labeling of activations in SPM using a macroscopic anatomical parcellation of the MNI MRI single-subject brain. Neuroimage **15**, 273 (2002)
36. Mensch, A., Mairal, J., Thirion, B., Varoquaux, G.: Stochastic subsampling for factorizing huge matrices. IEEE Trans. Signal Process. **66**(1), 113–128 (2018)
37. Dadi, K., et al.: Benchmarking functional connectome-based predictive models for resting-state fMRI. Neuroimage **192**, 115–134 (2019)
38. Fischl, B., et al.: Automatically parcellating the human cerebral cortex. Cereb. Cortex **14**(1), 11–22 (2004)
39. O'Donnell, L.J., et al.: Automated white matter fiber tract identification in patients with brain tumors. NeuroImage: Clin. **13**, 138–153 (2017)
40. Kumar, V., et al.: mulEEG: a multi-view representation learning on EEG signals. In: Medical Image Computing and Computer Assisted Intervention - MICCAI 2022, pp. 398–407 (2022)
41. Chang, L.J., Gianaros, P.J., Manuck, S.B., Krishnan, A., Wager, T.D.: A sensitive and specific neural signature for picture-induced negative affect. PLoS Biol. **13**, 1–28 (2015)
42. Carlson, D., et al.: Dynamically timed stimulation of corticolimbic circuitry activates a stress-compensatory pathway. Biol. Psychiatry **82**(12), 904–913 (2018)

43. Yamins, D.L., Hong, H., Cadieu, C.F., Solomon, E.A., Seibert, D., DiCarlo, J.J.: Performance-optimized hierarchical models predict neural responses in higher visual cortex. Proc. Natl. Acad. Sci. U.S.A. **111**(23), 8619–8624 (2014)

44. Deco, G., Jirsa, V.K., Robinson, P.A., Breakspear, M., Friston, K.J.: The dynamic brain: from spiking neurons to neural masses and cortical fields. PLoS Comput. Biol. **4**(8), e1000092 (2008)

45. Ratan Murty, N.A., Bashivan, P., Abate, A., DiCarlo, J.J., Kanwisher, N.: Computational models of category-selective brain regions enable high-throughput tests of selectivity. Nat. Commun. **12**(1), 1–14 (2021)

46. Liu, T.T.: Noise contributions to the FMRI signal: an overview. Neuroimage **143**, 141–151 (2016)

47. Power, J.D., Barnes, K.A., Snyder, A.Z., Schlaggar, B.L., Petersen, S.E.: Spurious but systematic correlations in functional connectivity MRI networks arise from subject motion. Neuroimage **59**(3), 2142–2154 (2012)

48. Laumann, T.O., et al.: On the stability of BOLD fMRI correlations. Cereb. Cortex **27**(10), 4719–4732 (2016)

49. Behzadi, Y., Restom, K., Liau, J., Liu, T.T.: A component based noise correction method (CompCor) for BOLD and perfusion based fMRI. Neuroimage **37**, 90 (2007)

50. Bijsterbosch, J.D., et al.: The relationship between spatial configuration and functional connectivity of brain regions. eLife **7**, 32992 (2018)

51. Gratton, C., et al.: Functional brain networks are dominated by stable group and individual factors, not cognitive or daily variation. Neuron **98**(2), 439–4525 (2018)

52. Nastase, S.A., Gazzola, V., Hasson, U., Keysers, C.: Measuring shared responses across subjects using intersubject correlation. Soc. Cognit. Affect. Neurosci. **14**(6), 667–685 (2019)

53. Robinson, E.C., et al.: MSM: a new flexible framework for multimodal surface matching. Neuroimage **100**, 414–426 (2014)

Data Science: Architectures

A Novel Feature Selection Based Text Classification Using Multi-layer ELM

Rajendra Kumar Roul[1](✉) ⓘD and Gaurav Satyanath[2] ⓘD

[1] Department of Computer Science, Thapar Institute of Engineering and Technology, Patiala, Punjab, India
raj.roul@thapar.edu
[2] Department of Electrical and Computer Engineering, Carnegie Mellon University, Pittsburgh, PA, USA
gsatyana@andrew.cmu.edu

Abstract. Deep learning architectures used for text classification are becoming increasingly prevalent. However, the existing deep architectures have flaws such as slow speed, long training times, and the local minimum problem. Multi-layer Extreme Learning Machine has overcome these problems by avoiding backpropagation and thus saves a significant amount of training time, ensures global optimal, and can handle a vast quantity of data. The most important characteristic of Multi-layer ELM is its *feature space (FS)*, which allows the input features to be linearly separated without using any kernel techniques. The architecture of Multi-layer ELM and its technique of feature mapping are examined in this research with the help of a novel feature selection technique termed as Correlation-based Feature Selection (*CORFS*). Empirical results of the proposed feature selection technique are compared with state-of-the-art techniques. Different classification algorithms are extensively tested on Multi-layer ELM feature space and on TFIDF vector space to demonstrate the efficiency of the feature mapping technique. Results of the experiment revealed that the proposed feature selection technique is better than the conventional feature selection techniques, and the feature space of Multi-layer ELM outperforms TFIDF.

Keywords: Classification · Deep network · Multi-layer ELM · TF-IDF · Vector space

1 Introduction

Text mining can be understood as data mining on textual documents. Typical text mining tasks are text classification, clustering, retrieval, etc. Most of the earlier works used traditional machine learning techniques for text classification, such as support vector machine, naive Bayes, logistic regression, maximum entropy, decision trees, etc. But they are not able to capture the discriminative features automatically from the training data. Their performances heavily depend on data representation, and it is labor-intensive. The literature on text classification has been dominated by deep learning techniques motivated by the outstanding results of deep neural networks in text mining, image processing, and natural language processing [1,2]. But they have limitations like they need large memory bandwidth, huge training time is required because

P. P. Roy et al. (Eds.): BDA 2022 India, LNCS 13773, pp. 33–52, 2022.
https://doi.org/10.1007/978-3-031-24094-2_3

of backpropagation, architecture is very complex, preserving interdependencies among the internal layers for a long time is quite difficult etc. Hence it is not easy to generalize the text classification models to a new domain. An efficient deep learning classifier called Multi-layer ELM was introduced in the year 2013 by Kasun et al. [3] to address the above problems.

1.1 Research Motivation

Overall these existing machine and deep classification techniques have the following limitations:

i. Displaying the data becomes more complex when a large storage space is required due to the growth of the dataset size.
ii. When the input data grows exponentially on the limited dimensional space, distinguishing input features on*TF-IDF* vector space becomes challenging.
iii. Machine and deep learning classifier performances heavily depend on data representation, which is labor-intensive.

Feature selection which selects an optimal subset from the massive volume of the dataset, can alleviate the dimensionality curse but cannot separate the features in lower dimensional space due to the dynamic growth of data items [4,5]. Kernel approaches [6,7] are commonly utilized by any classification process to deal with this challenge. Kernel methods have been used for classification techniques in the past, and better results have been obtained. A detailed survey of kernel and spectral methods for classification has been done by Filippone M et al. [8]. Though kernel methods can handle the data separation in a lower dimensional space by projecting them to a higher-dimensional space, they are expensive (i.e., time-consuming) because of using the dot product to compute the structural similarity among the input features. Using the feature mapping technique of ELM, Huang et al. [9] admitted that by mapping the input vector non-linearly to a high-dimensional feature space, the features become simple and separable linearly, and thus can outperform the kernel approaches [10]. But ELM is a single-layer architecture, thus requiring an extensive network, which is challenging to design to perfectly match the heavily changed input data.

In this vein, this research investigated the feature space of Multi-layer ELM (ML-ELM) [11], which extensively exploits the advantages of *ELM feature mapping* [12,13] and ELM autoencoder to address the constraints mentioned above. The goal of this study is to investigate the extended feature space of ML-ELM (*HDFS-MLELM*) and to thoroughly test this feature space for text classification in comparison to the *TF-IDF vector space* (*VS-TFIDF*).

1.2 Research Contribution

The major contributions of the paper can be summarized as follows:

- This work studies *HDFS-MLELM*, and uses text data to thoroughly investigate multiple classification algorithms on *HDFS-MLELM* and on *VS-TFIDF*.

- It is clear from the past literature that no research on classification using text data has been done on the Multi-layer ELM's enlarged feature space. As a result, in light of the benefits mentioned above, this study can be considered as a new direction in the text classification domain.
- A novel feature selection termed Correlation-based Feature Selection *CORFS* is proposed for selecting the essential features from a big corpus.
- To demonstrate its usefulness, the performance of Multi-layer ELM employing the suggested *CORFS* technique has been compared with several machines and deep learning classifiers.
- Text classification results of various traditional classifiers after running them on ELM feature space and on *HDFS-MLELM* are compared in order to show the effectiveness of *HDFS-MLELM*.
- The experimental results of the proposed approach are compared with the state-of-the-art approaches.

Rest of the paper is as follows: Sect. 2 introduces the preliminaries of Multi-layer ELM and its feature mapping technique. The proposed methodology is discussed in Sect. 3. Section 4 carried out the experimental work. The paper is concluded in Sect. 5.

2   Prelims

2.1   Multi-layer ELM

As demonstrated in Fig. 3, Multi-layer ELM (ML-ELM) is a hybrid of ELM (shown in Fig. 1) and ELM autoencoder (shown in Fig. 2) with more than one hidden layer and is discussed using the following steps.

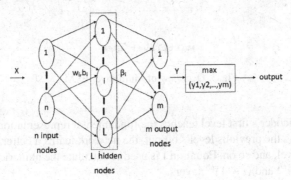

Fig. 1. Overview of ELM

- Unsupervised training occurs between the hidden layers using ELM Autoencoder [14]. Unlike other deep networks, ML-ELM does not require fine-tuning since ELM's autoencoder capacity is an excellent match for ML-ELM [15].
- Stacks are built on top of the ELM Autoencoder in a progressive way to create a multi-layer neural network architecture. The output of one trained ELM Autoencoder is fed into the next ELM Autoencoder, and so on.

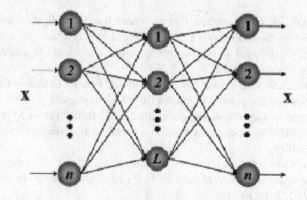

Fig. 2. Overview of ELM-autoencoder

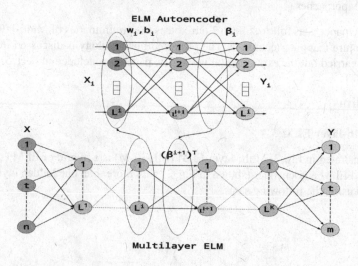

Fig. 3. Overview of Multi-layer ELM

- ELM Autoencoder's first level teaches the fundamental representation of input data. By integrating the previous level's output, the network learns a better representation in the next level, and so on. Equation 1 is used to calculate the numerical understanding between i^{th} and $(i-1)^{th}$ layers.

$$H_i = g((\beta_i)^T H_{i-1}) \tag{1}$$

where H_{i-1} and H_i are the input and output matrices of the ith hidden layer, respectively. $g(.)$ is the activation function, and β is the learning parameter. The input layer is H_0, and the first hidden layer is H_1. Regularized least squares is used to get the output weight β [16].
- Finally, supervised learning is utilized to fine-tune the network (ELM is used for this purpose).

3   Methodology

1. *Documents Pre-processing*:
   Let corpus P consists of C classes. At the beginning of the feature engineering, all documents of each class are combined into a single set called D_{large}. Then lexical-analysis, stop-word deletion, HTML tag removal, and stemming [1] are done on D_{large}. Natural Language Toolkit[2] is used to extract index terms from d_{large}. After completing the basic data cleaning, the first set of features was derived from D_{large} and created a term-document matrix.

2. *Correlation Based Feature Selection (CORFS)*:
   Using k-means[3] clustering algorithm [17], the D_{large} is divided into n term-document clusters $td_i, i \in [1, n]$. The following steps discuss the methodology used to extract important features from each cluster td_i.

   i. Calculating Centroid:
      First the centroid of td_i is calculated using Eq. 2.

      $$sc_i = \frac{\sum\limits_{j=1}^{r} t_i}{r} \tag{2}$$

      Then cosine-similarity is computed between $t_j \in td_i$ and sc_i.

   ii. Generating correlation matrix:
      Equation 3 is used to find the correlation (cr)[4] between pair of terms t_i and t_j and is shown in Table 1.

      $$cr_{t_i t_j} = \frac{C_{t_i t_j}}{\sqrt{(V_{t_i} * V_{t_j})}} \tag{3}$$

      where, $C_{t_i t_j}$ is the covariance (joint variability between two terms) between t_i and t_j. V_{t_i} and V_{t_j} are their variances respectively as defined below.

      $$V_{t_i} = \frac{1}{b-1} \sum_{m=1}^{b} (X_{im} - \overline{X_i})^2$$

      $$V_{t_j} = \frac{1}{b-1} \sum_{m=1}^{b} (X_{jm} - \overline{X_j})^2$$

      where $\overline{X_i}$ and $\overline{X_j}$ represents the mean of b documents having the terms t_i and t_j respectively. The covariance between t_i and t_j is computed using Eq. 4.

      $$C_{t_i t_j} = \frac{1}{b-1} \sum_{m=1}^{b} (X_{im} - \overline{X_i})(X_{jm} - \overline{X_j}) \tag{4}$$

---

[1] https://pythonprogramming.net/lemmatizing-nltk-tutorial/.
[2] https://www.nltk.org/.
[3] k value is decided based on the experiment for which the best result is obtained.
[4] https://libguides.library.kent.edu/SPSS/PearsonCorr.

Table 1. Correlation matrix

	t_1	t_2	t_3	$\cdots$	t_r
t_1	cr_{11}	cr_{12}	cr_{13}	$\cdots$	cr_{1r}
t_2	cr_{21}	cr_{22}	cr_{23}	$\cdots$	cr_{2r}
t_3	cr_{31}	cr_{32}	cr_{33}	$\cdots$	cr_{3r}
$\vdots$	$\vdots$	$\vdots$	$\vdots$	$\ddots$	$\vdots$
t_r	cr_{r1}	cr_{r2}	cr_{r3}	$\cdots$	cr_{rr}

iii. Rejection of high correlated terms from td_i:

Terms that are highly correlated in a cluster are generally considered as a sort of synonym, and hence they do not discriminate well in the cluster. Therefore, those terms should be removed from the cluster. To find those terms in td_i, initially, those terms that have the maximum cosine-similarity score in td_i get selected. Subsequently, a set of terms are identified which are highly correlated to t_i (≤ -0.87 or ≥ 0.89)[5] and that set of terms get removed from td_i. This step is repeated for the next highest cosine-similarity score term and so on till td_i gets exhausted. Finally, all highly correlated terms are removed from td_i.

iv. Computing Discriminating Power Measure (*DPM*):

(*DPM*) [18] is a technique that measures the relevance, i.e., the importance of a term in a cluster. If the *DPM* score of a term inside an unbiased cluster is very high, then that term is an important term for that cluster. It is because many documents of the cluster contain that term. The cohesion or tightness of that term is very close to the cluster's center.

– For each $t_i \in td_i$, the document frequency inside (DF_{in,t_i}) and outside (DF_{out,t_i}) of td_i are calculated using Eqs. 5 and 6 respectively.

$$DF_{in,t_i} = \frac{no.\ of\ documents\ \in td_i\ and\ have\ t_i}{no.\ of\ documents\ \in td_i} \tag{5}$$

$$DF_{out,t_i} = \frac{no.\ of\ documents\ have\ t_i\ and\ \notin td_i}{no.\ of\ documents\ \notin td_i} \tag{6}$$

– The difference between inside and outside document frequency of $t_i \in td_i$ is computed using Eq. 7.

$$DIFF_{td_i,t_i} = |DF_{in,t_i} - DF_{out,t_i}| \tag{7}$$

– Equation 8 computes the *DPM* score of each term.

$$DPM(td_i, t_i) = \sum_{i=1}^{P} DIFF_{td_i,t_i} \tag{8}$$

v. Selection of candidate terms having High *DPM* scores:

of term-document cluster are arranged as per the *DPM* scores, and higher $k\%$ terms are selected as the candidate terms. This step is repeated for each td_i so that every td_i has top $k\%$ candidate terms in them.

---

[5] decided experimentally so that we will not lose more terms.

3. Input feature vector generation:
   To build the input feature vector, all the top $k\%$ features of each td_i are merged into a list L_{list}.
4. Feature mapping of Multi-layer ELM:
   i. Multi-layer ELM heavily employs the universal classification [19,20] and approximation [21,22] capabilities of ELM.
   ii. ML-ELM cleverly leveraged the extended representation (i.e., $n < L$) technique of the ELM autoencoder [12,23], where n and L are the number of input and hidden layer nodes, respectively.
   iii. The features of ML-ELM are transferred from a low-dimensional feature space to a higher-dimensional feature space using Eq. 9. Mapping of the input vector to *HDFS-MLELM* is shown in Fig. 4 where, $h_i(\mathbf{x}) = g(w_i.\mathbf{x} + b_i)$.

$$
h(\mathbf{x}) =
\begin{bmatrix}
h_1(\mathbf{x}) \\
h_2(\mathbf{x}) \\
h_3(\mathbf{x}) \\
. \\
. \\
. \\
h_L(\mathbf{x})
\end{bmatrix}^T
=
\begin{bmatrix}
g(w_1, b_1, \mathbf{x}) \\
g(w_2, b_2, \mathbf{x}) \\
g(w_3, b_3, \mathbf{x}) \\
. \\
. \\
. \\
g(w_L, b_L, \mathbf{x})
\end{bmatrix}^T
\tag{9}
$$

$h(\mathbf{x}) = [h_1(\mathbf{x}), h_2(\mathbf{x}), \cdots, h_i(\mathbf{x}), \cdots, h_L(\mathbf{x})]^T$ transfer the input features to *HDFS-MLELM* [24,25].

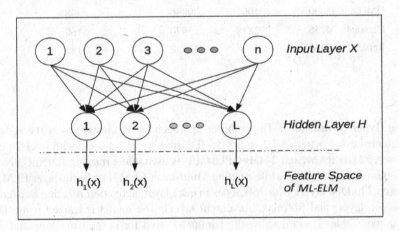

Fig. 4. Feature mapping technique of ML-ELM

   iv. L_{list} is mapped into *MLELM-HDFS* using Eq. 9. Before the transformation, L is set to a higher value than n. This makes all the features of L_{list} linearly separable.
5. Classification on *MLELM-HDFS*:
   Different supervised learning algorithms employing L_{list} as the input feature vector are run individually on *TFIDF-VS* and *MLELM-HDFS* respectively.

4   Analysis of Experimental Results

The setup for the experimental study is detailed in depth in this section. The performance evaluation of state-of-the-art classification algorithms in the feature space of ML-ELM is examined thoroughly. Experiments were done on the feature space of ML-ELM by altering the number of hidden layer nodes L of ML-ELM as per the three representations mentioned below, where n is the number of nodes in the input layer.

- for compress representation ($n > L$): $L = 0.4n$ and $L = 0.7n$
- for extended representation ($n < L$): $L = 1.4n$ and $L = 1.2n$
- for equal representation: ($n = L$): $L = 1.0n$

4.1   Experimental Setup

A Brief Description of the Datasets Utilized in the Experiment: To conduct the experiment, four benchmark datasets (WebKB[6], Classic4[7], 20-Newsgroups[8], and Reuters[9] are used and the details are shown in Table 2.

Table 2. Corpus statistics

Datasets	Training docs	Testing docs	Terms used for training	10% of terms
20-NG	11292	7527	32269	3239
DMOZ	38000	31067	39886	3989
Classic4	4256	2838	15970	1602
Reuters	5484	2188	13532	1351

Tuning Hyper-parameters: The proposed approach for the classification of text data is implemented using python 3.7.3 on Spyder IDE running on a system with Intel Core i11 processor, 32 GB RAM, and 24 GB GPU. GPU is used while running ANN, CNN, and RNN algorithms, and CPU while running Multi-layer ELM. For Multi-layer ELM, we have used 3 hidden layers with 150 nodes in each layer, activation function as Sigmoid (for hidden layer) and Softmax (for output layer). The model is trained using DGX workstation. Tables 3 and 4 show the parameter used for several machine and deep learning algorithms, respectively. Fixing all parameter values is done by repeating the experiment.

---

[6] http://www.cs.cmu.edu/afs/cs/project/theo-20/www/data/.

[7] http://www.dataminingresearch.com/index.php/2010/09/classic3-classic4-datasets/.

[8] http://qwone.com/~jason/20Newsgroups/.

[9] http://www.daviddlewis.com/resources/testcollections/reuters21578/.

Table 3. Setting different parameters (machine learning)

Classifier	Tuned parameter
SVM	kernel: {'linear'}, random_state: {0}, C: {0.025,1}, gamma: {1}, degree: {3}
K-NN	k: {5}, euclidean distance: {2}
Decision Trees	min_samples_leaf: {5}, criterion: {'entropy'}, random_state: {0}, max_depth: {10, 150, 500}
Random Forest	random_state: {0}, bootstrap:{'True'}, criterion: {'entropy'}, max_depth: {3, 150, 500, 1000}, n_estimators: {100}
Naive Bayes	alpha: {1}, fit_prior: {True}, class_prior: {None}, binarize: {0}
Extra Trees	max_depth: {3}, criterion: {'entropy'}, n_estimators: {100}, random_state: { 0}, min_samples_split: {5}, min_samples_leaf: {5}, max_features: {50}
ELM	no. of hidden layer: {1}, no. of nodes in the hidden layer: {150}, activation function: hidden layer({sigmoid}), ouptput layer ({softmax})
Adaboost	subsample: {0.5}, n_estimators: {10}, learning_rate: {1}, random_state: {0}, max_depth: {5},
Gradient Boosting	min_samples_split: {2}, min_samples_leaf: {5}, learning_rate:{ 0.1(shrinkage)}, subsample: {0.4}, random_state: {0}, n_estimators: {75 (no. of trees)}, max_depth: {3}

Table 4. Setting different parameters (deep learning)

Classifier	No. of Hidden layers	Activation-function	Dropout	Optimizer	Epoch	Batch size
CNN	3	ReLU(hidden layer), Softmax(output layer)	0.5	ADAM	190	80
ANN	4	Sigmoid (hidden layer), Sigmoid(output layer)	0.3	SGD	250	120
RNN	3	Tanh (hidden layer), Softmax(output layer)	0.3	GD	220	110

4.2 Discussion

Performance Evaluation of *CORFS* Technique: The proposed *CORFS* technique is compared with different traditional feature selection techniques (Bi-normal separation (BNS), Mutual Information (MI), Chi-square, and Information Gain (IG)), and the F-measures are shown in Tables 5, 6, 7, 8, 9, 10, 11, 12, 13, 14, 15 and 16 respectively for different datasets on top 1%, 5%, and 10% features, where bold indicates maximum. The F-measure of the proposed *CORFS* approach is compared with the state-of-the approaches, which is summarized in Table 17. The findings suggest that the proposed feature selection approach is equivalent to or better than the previous one and can be used to classify text documents using the ML-ELM feature space.

Performance Comparisons of Multi-layer ELM: It's worth noting that ML-ELM outperforms other machine learning classifiers in most feature selection strategies across various datasets, as shown in Table 18. Figures 5 and 6 show F-measure and accuracy comparisons of Multi-layer ELM with various deep learning techniques using the *CORFS* approach. Results indicate the effectiveness of ML-ELM over the machine and deep learning classifiers.

Table 5. 20-NG (Top 1%)

Classifier	BNS	Chi-square	IG	MI	CORFS
Linear SVC	0.89121	0.87162	0.87361	0.88946	0.87514
SVM	0.89233	0.88966	0.89152	0.89443	0.89427
Decision Trees	0.88226	0.86306	0.86991	0.87513	0.87781
Gradient Boosting	0.85582	0.83783	0.84083	0.86062	0.85223
Adaboost	0.87321	0.88313	0.88382	0.88471	0.87334
NB(Multinomial)	0.86302	0.83861	0.83794	0.85793	**0.88661**
ML-ELM	0.91702	0.91238	0.90512	0.91669	**0.93803**
ELM	0.89241	0.87042	0.87544	0.89572	0.88421
Extra Trees	0.89671	0.87605	0.88642	0.88234	0.88761
RF	0.85992	0.85846	0.85901	0.84242	0.85584

Table 6. 20-NG (Top 5%)

Classifier	BNS	Chi-square	IG	MI	CORFS
Linear SVC	0.94377	0.93461	0.93729	0.93375	**0.95509**
SVM	0.93459	0.92816	0.92814	0.93014	0.93457
Decision Trees	0.93516	0.88816	0.90252	0.92878	0.93314
Gradient Boosting	0.89954	0.88561	0.89489	0.89591	0.87908
Adaboost	0.89075	0.88366	0.86261	0.87112	0.87184
NB(Multinomial)	0.93122	0.90103	0.91607	0.92511	0.92191
ML-ELM	0.93873	0.94882	0.94664	0.95584	**0.96746**
ELM	0.93551	0.92673	0.93012	0.93682	0.92331
Extra Trees	0.90426	0.88001	0.90223	0.89423	0.89717
RF	0.85997	0.86254	0.85813	0.85763	0.85618

Table 7. 20-NG (Top 10%)

Classifier	BNS	Chi-square	IG	MI	CORFS
Linear SVC	0.94741	0.93731	0.93647	0.94376	**0.94922**
SVM	0.94284	0.94557	0.93646	0.94652	0.94537
Decision Trees	0.93995	0.91011	0.93352	0.93991	0.91132
Gradient Boosting	0.90146	0.89581	0.89862	0.90511	0.89581
Adaboost	0.88265	0.86262	0.86342	0.87252	0.87685
NB(Multinomial)	0.93821	0.92343	0.93271	0.93731	0.91936
ML-ELM	0.95635	0.96881	0.95566	0.95813	**0.96927**
ELM	0.93228	0.94052	0.93442	0.94671	0.92885
Extra Trees	0.89292	0.89241	0.89471	0.89272	**0.90571**
RF	0.86371	0.84922	0.86604	0.85912	0.85642

Table 8. Classic4 (Top 1%)

Classifier	BNS	Chi-square	IG	MI	CORFS
Linear SVC	0.92632	0.90663	0.92185	0.92799	0.91670
SVM	0.91492	0.88287	0.89943	0.91785	0.90146
Decision Trees	0.83281	0.79881	0.87911	0.86601	0.83021
Gradient Boosting	0.90966	0.83336	0.88875	0.88188	0.88345
Adaboost	0.89162	0.88094	0.88437	0.88986	0.88393
NB(Multinomial)	0.84197	0.76852	0.80705	0.85317	**0.88163**
ML-ELM	0.94651	0.92222	0.91885	0.94563	**0.95707**
ELM	0.90285	0.88142	0.89945	0.92383	0.90181
Extra Trees	0.91773	0.89212	0.91534	0.91801	0.88714
RF	0.86046	0.84001	0.85312	0.86365	0.85874

Table 9. Classic4 (Top 5%)

Classifier	BNS	Chi-square	IG	MI	CORFS
Linear SVC	0.94595	0.92332	0.94104	0.94301	**0.95654**
SVM	0.96512	0.93917	0.94051	0.94793	0.96022
Decision Trees	0.92631	0.90762	0.91491	0.90219	0.90988
Gradient Boosting	0.93411	0.92597	0.92951	0.93952	0.93277
Adaboost	0.87344	0.88183	0.84972	0.85483	0.84584
NB(Multinomial)	0.93813	0.92326	0.93096	0.94524	**0.94666**
ML-ELM	0.96667	0.94887	0.96254	0.96547	0.96542
ELM	0.94531	0.92523	0.94675	0.94171	0.94577
Extra Trees	0.91892	0.91892	0.91657	0.91922	0.89553
RF	0.84924	0.84847	0.84886	0.85558	0.84848

Table 10. Classic4 (Top 10%)

Classifier	BNS	Chi-square	IG	MI	CORFS
Linear SVC	0.94542	0.92652	0.92758	0.97287	0.96768
SVM	0.94546	0.92124	0.92339	0.96868	0.96552
Decision Trees	0.92031	0.90688	0.91548	0.91054	0.89722
Gradient Boosting	0.94021	0.93594	0.93951	0.94238	0.93991
Adaboost	0.85482	0.85482	0.85482	0.84587	0.84587
NB(Multinomial)	0.95097	0.94561	0.94778	0.95352	**0.96918**
ML-ELM	0.97106	0.95877	0.96886	0.97944	**0.98577**
ELM	0.92333	0.94672	0.95634	0.96948	0.95633
Extra Trees	0.91525	0.92688	0.91851	0.91436	0.91675
RF	0.85072	0.84735	0.85205	0.84573	0.85087

Reasons for Better Performance of Multi-layer ELM over Other Classifiers: The following points highlighted the basic reasons behind the superiority of ML-ELM.

i. In ML-ELM, there is no need to fine-tune the hidden node settings and other parameters, and no back-propagations are required. This saves training time, and the learning speed becomes exceedingly rapid throughout the classification phase.

ii. ML-ELM is less expensive than other deep learning architectures because it does not require any GPU to run. When the dataset size grows, excellent performance is realized in ML-ELM.

iii. ML-ELM can map and linearly separate a huge volume of data in the extended space, thanks to its universal approximation and classification capabilities.

iv. The training in ML-ELM is mostly unsupervised except at the last level, where it is supervised.

v. Multiple hidden layers provide a high-level data abstraction, and each layer learns new input forms, making ML-ELM more efficient.

Table 11. Reuters (Top 1%)

Classifier	BNS	Chi-square	IG	MI	CORFS
Linear SVC	0.93365	0.92335	0.92966	0.93972	0.91524
SVM	0.93242	0.93918	0.93146	0.94952	0.94924
Decision Trees	0.86441	0.85532	0.85344	0.85343	0.85147
Gradient Boosting	0.83336	0.83041	0.83153	0.83832	0.83452
Adaboost	0.64006	0.64004	0.64002	0.76296	**0.77929**
NB(Multinomial)	0.87202	0.84181	0.85837	0.86014	**0.87526**
ML-ELM	0.95413	0.95672	0.95679	0.96758	0.94602
ELM	0.94566	0.95022	0.93142	0.93375	0.93326
Extra Trees	0.92232	0.92944	0.92361	0.93203	0.90924
RF	0.89168	0.88852	0.89003	0.89517	0.88447

Table 12. Reuters (Top 5%)

Classifier	BNS	Chi-square	IG	MI	CORFS
Linear SVC	0.95122	0.94781	0.94007	0.95424	0.94074
SVM	0.95228	0.94387	0.94643	0.96554	**0.96889**
Decision Trees	0.82963	0.87127	0.85762	0.85328	0.84876
Gradient Boosting	0.84274	0.83727	0.84063	0.84358	0.83985
Adaboost	0.63822	0.65845	0.62866	0.67425	**0.73312**
NB(Multinomial)	0.90244	0.89552	0.90328	0.91178	0.90667
ML-ELM	0.96395	0.96317	0.95839	0.96878	**0.96938**
ELM	0.94569	0.94897	0.95605	0.93452	0.94452
Extra Trees	0.92823	0.92323	0.93327	0.91607	0.91752
RF	0.90287	0.90607	0.90373	0.90356	0.89664

Table 13. Reuters (Top 10%)

Classifier	BNS	Chi-square	IG	MI	CORFS
Linear SVC	0.95735	0.94174	0.94453	0.94695	0.95477
SVM	0.95487	0.94617	0.94684	0.94818	**0.96782**
Decision Trees	0.79521	0.84727	0.83482	0.81192	0.79144
Gradient Boosting	0.84552	0.84322	0.84262	0.84547	0.84518
Adaboost	0.63822	0.63841	0.63427	0.63416	**0.64808**
NB (Multinomial)	0.90078	0.90556	0.90812	0.90972	0.88692
ML-ELM	0.96723	0.95765	0.96772	0.96322	**0.96957**
ELM	0.92334	0.94666	0.94549	0.95504	0.94558
Extra Trees	0.91903	0.91693	0.91904	0.91982	0.91077
RF	0.90554	0.89852	0.90683	0.89948	0.90657

Table 14. DMOZ (Top 1%)

Classifier	BNS	Chi-square	IG	MI	CORFS
Linear SVC	0.84016	0.81564	0.83202	0.77058	0.81838
SVM	0.85637	0.82417	0.84035	0.81685	**0.87784**
Decision Trees	0.65441	0.63747	0.67972	0.53565	0.63932
Gradient Boosting	0.76795	0.75208	0.76272	0.74438	0.76457
Adaboost	0.83837	0.81692	0.81552	0.79221	**0.84301**
NB(Multinomial)	0.68786	0.65052	0.66733	0.61532	0.63171
ML-ELM	0.87502	0.85369	0.86633	0.83656	0.84518
ELM	0.81261	0.80381	0.84786	0.78943	0.80537
Extra Trees	0.84671	0.82841	0.83191	0.81561	0.81501
RF	0.79123	0.77397	0.77501	0.74931	0.76596

Table 15. DMOZ (Top 5%)

Classifier	BNS	Chi-square	IG	MI	CORFS
Linear SVC	0.87807	0.86862	0.87753	0.88152	0.86491
SVM	0.87913	0.88137	0.88866	0.88722	0.85949
Decision Trees	0.70749	0.74333	0.73445	0.70754	0.71887
Gradient Boosting	0.79423	0.78232	0.78506	0.78552	0.77723
Adaboost	0.79832	0.81162	0.81848	0.80972	**0.82283**
NB(Multinomial)	0.77866	0.75184	0.76398	0.75469	0.76444
ML-ELM	0.89678	0.88565	0.90039	0.90672	**0.90918**
ELM	0.86724	0.83253	0.84516	0.87521	0.84679
Extra Trees	0.83864	0.82967	0.84824	0.84483	0.83079
RF	0.78484	0.78763	0.79159	0.78826	0.77707

Table 16. DMOZ (Top 10%)

Classifier	BNS	Chi-square	IG	MI	CORFS
Linear SVC	0.88881	0.88006	0.86353	0.88186	0.87572
SVM	0.87053	0.88834	0.86881	0.88044	0.87601
Decision Trees	0.70036	0.73239	0.71893	0.70537	0.70294
Gradient Boosting	0.80386	0.79181	0.79742	0.79183	0.78614
Adaboost	0.81145	0.79832	0.79835	0.80098	0.80889
NB(Multinomial)	0.78502	0.77861	0.78698	0.77904	**0.79622**
ML-ELM	0.90231	0.90438	0.90533	0.90988	**0.91279**
ELM	0.86626	0.88769	0.88013	0.87566	0.84248
Extra Trees	0.83462	0.83173	0.83147	0.82881	0.81986
RF	0.78082	0.79575	0.79054	0.78477	0.77781

Performance Evaluation of Classification Algorithms: For practical reasons, six distinct classification approaches are performed on the *HDFS-MLELM* and the *VS-TFIDF*, employing four datasets individually. The obtained accuracies and F-measures are shown in Figs. 7, 8, 9 and 10 and Figs. 11, 12, 13 and 14 respectively.

The following conclusions are drawn from the findings:

i. Compared to the *VS-TFIDF*, the empirical findings in all three feature spaces of Multi-layer ELM are superior.
ii. Linear SVM outperforms other supervised learning algorithms, owing to its convex optimization property [35] and generalization property [36], both of which are independent of feature space dimension.
iii. F-measure and accuracy are better in *HDFC-MLELM* whereas it is close on equal dimensional space.

The performance of the proposed approach is compared with the state-of-the-art classification approaches, and the results are shown in Table 19, where bold indicates the maximum accuracy.

Table 17. Performance of Feature selection algorithms (bold indicates maximum)

Authors	Classifier used	Dataset	F-measure (%)
Wang et al. [26]	KNN, Naive-Bayes, Decision Trees, SVM, Random Forest, Booster **Best performance: Booster**	New Popularity	67.31
Khoder et al. [27]	KNN, SVM, LDA, ICS_DLSR, EM_ICS_FS **Best performance: EM_ICS_FS**	Extended Yale B Face, Outdoor Scene	95.88
Stefano et al. [28]	Naive-Bayes, KNN, SVM, Multi-layer perceptron (MLP), **Best performance: MLP**	OPTODIGIT, MFEAT, MNIST, NIST	95.07
Tubishat et al. [29]	Association Rule Mining (ARM), Meta classifier, **Best performance: Meta classifier**	Opinion Corpus for Arabic (OCA)	90.80
Jiang et al. [30]	MDEFS, MDEFS(NE), BBPSO-FS, **Best performance: BBPSO-FS**	LSVT, Waveform1	91.78
Got et al. [31]	FW, BDE, BPSO, BGWO, WOA, jDE **Best performance:FW**	Musk, Sobar, Spectf	88.97
Miao et al. [32]	LLfea, Lapscore, SPEC, MCFS, UDFS, EUFS, RSR, NOVRSR, **Best performance:NOVRSR**	BA, JAFFE, ORL	52.18
Ezenkwu et al. [33]	Random Forest, SVM, **Best performance: Random Forest**	Ionosphere, Glass Identification, Dermatology, Isolet, Statlog Heart, Landsat satellite, Semeion, Soybean	70.43
Adamu et al. [34]	CSA, CCSA, PSO, BPSO, ECCSPSOA, **Best performance: ECCSPSOA**	Wine, Dermatology, Heart, Ionosphere, Lung Cancer, Hepatitis, Parkinson, Divorce, Thoracic Surgery, Phishing Website, Absenteeism at Work	89.76
Proposed (*CORFS*)	Linear SVC, SVM, Decision Trees, Gradient Boosting, Adaboost, Multinomial NB, ML-ELM, ELM, Extra trees, Random Forest, **Best performance: ML-ELM**	20-NG, Classic4, Reuters, DMOZ	**98.57**

Table 18. Comparing ML-ELM with machine learning classifiers using *CORFS*

Classifier	20- Newsgroups			Classic4			Reuters			DMOZ		
	1%	5%	10%	1%	5%	10%	1%	5%	10%	1%	5%	10%
SVC (linear)	87.514	93.454	94.921	90.146	95.655	96.551	91.524	94.074	95.477	81.838	86.491	87.572
SVM (linear)	89.428	95.509	94.536	91.670	96.021	96.768	**94.924**	96.889	96.782	**87.784**	85.949	87.601
Gradient Boosting	85.224	87.908	89.582	88.345	93.278	93.991	83.452	83.985	84.518	76.457	77.723	78.614
Decision Trees	87.782	93.314	91.133	83.021	90.987	89.722	85.147	84.876	79.144	63.932	71.887	70.294
NB (Multinomial)	88.662	92.191	91.937	88.163	94.667	96.918	87.526	90.667	88.692	63.171	76.444	79.622
Adaboost	87.335	87.184	87.686	88.393	84.585	84.587	77.929	73.312	64.808	84.301	82.283	80.889
Random Forest	85.584	85.618	85.641	85.874	84.849	85.087	88.447	89.664	90.657	76.596	77.707	77.781
Extra Trees	88.762	89.717	90.572	88.714	89.554	91.675	90.924	91.752	91.077	81.501	83.079	81.986
ELM	88.420	92.331	92.886	90.182	94.578	95.633	93.326	94.453	94.558	80.537	84.679	84.248
MLELM	**93.802**	**96.746**	**96.928**	**95.707**	**96.541**	**98.577**	94.602	**96.938**	**96.957**	84.518	**90.918**	**91.279**

4.3 Comparisons of ELM and ML-ELM Feature Space

Traditional classifiers are run on *HDFS-MLELM* and ELM feature space. Figures 15, 16, 17 and 18 compare the performances of different classifiers on the higher dimensional feature space($L = 1.4n$) ML-ELM and ELM. The results indicate that the performances of classifiers are better in ML-ELM feature space compared to ELM feature space. The reason is due to the multilayer processing of ML-ELM compared to a single layer in ELM. SVM shows a better performance compared to other classifiers on both feature spaces.

Table 19. Performance of text classification algorithms (bold indicates maximum)

Authors	Classifier used	Dataset	Accuracy (%)
Guangquan *et al.* [37]	CNN, LSTM, **Best performance: CNN**	Amazon review dataset	97.33
Jeow *et al.* [38]	Random Forest, LR, LSTM, **Best performance: LSTM**	Notes dataset of Cincinnati Hospital Medical Centre	85.80
Hamouda *et al.* [39]	Naive-Bayes, Random Forest, SVM, LightGBM, Decision Trees, k-NN, **Best performance: SVM**	Arabic dataset	90.47
Bichitrananda *et al.* [40]	DNN, k-NN, SVM, RNN, CNN, FRS - RNN+ CNN, **Best performance: FRS-RNN + CNN**	20-Newsgroup	98.50
Janani *et al.* [41]	Naïve Bayes, k-NN, SVM, PNN, Adaboost, Random Forest, **Best performance: PNN**	20-Newsgroup, Reuters	93.70
Yan *et al.* [42]	k-NN, Decision Trees, Adaboost, FNN, SVM, HSAN-Capsule model, **Best performance: HSAN-Capsule**	Movie reviews dataset(IMDB)	90.12
Xiang *et al.* [43]	ABLSTM	online consultation data of medical healthcare	98.34
Shiyao *et al.* [44]	Frog-GNN	Amazon dataset, HuffPost dataset, FewRel dataset	94.28
Zhong *et al.* [45]	Multinomial Naive-bayes, SVM, k-NN, Decision Trees, Random Forest, Extra Trees, **Best performance: SVM**	Reuters-21578, 20 Newsgroups dataset	97.20
Shenghong *et al.* [46]	SVM, Neural network, Decision trees, Random Forest, Adaboost, **Best performance: SVM**	Chinese text dataset	79.50
Proposed work	Multinomial Naive-Bayes, Random Forest, k-NN, Linear SVM, Decision Trees, Extra Trees, **Best performance: Linear SVM on ML-ELM feature space**	20-NG, Classic4, Reuters, DMOZ	**98.80**

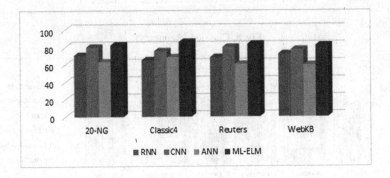

Fig. 5. F1-measure

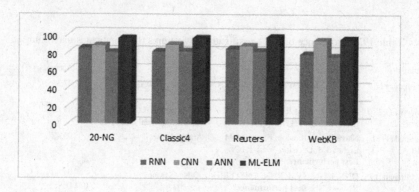

Fig. 6. Accuracy

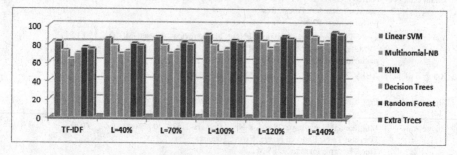

Fig. 7. 20-NG (Accuracy)

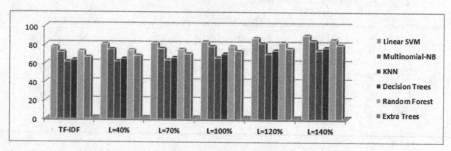

Fig. 8. Classic4 (Accuracy)

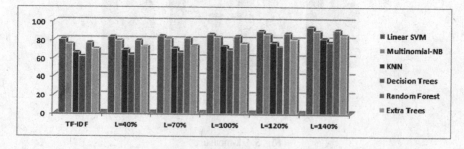

Fig. 9. Reuters (Accuracy)

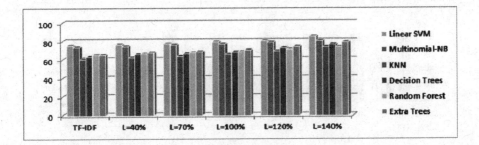

Fig. 10. DMOZ (Accuracy)

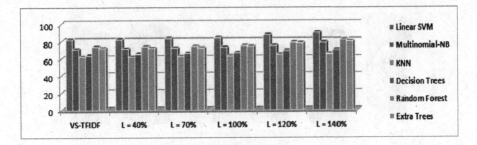

Fig. 11. 20NG (F1-measure)

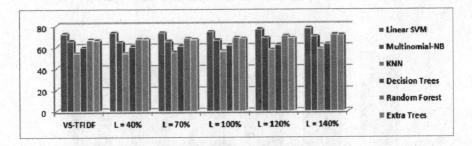

Fig. 12. Classic-4 (F1-measure)

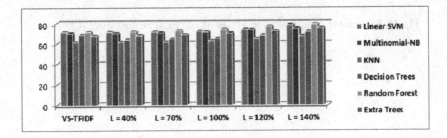

Fig. 13. Reuter (F1-measure)

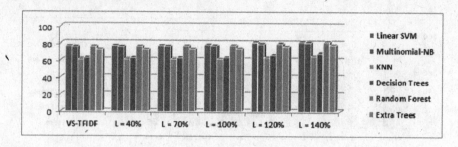

Fig. 14. DMOZ (F1-measure)

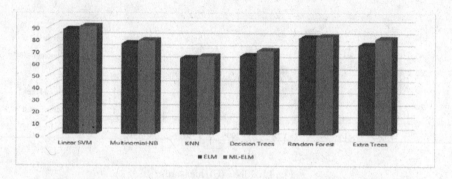

Fig. 15. F1-measure comparisons on ML-ELM and ELM Feature space (20-NG)

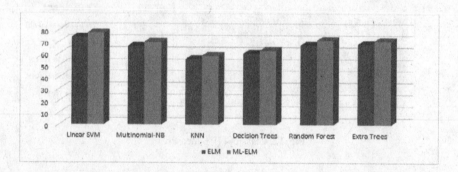

Fig. 16. F1-measure comparisons on ML-ELM and ELM Feature space (Classic4)

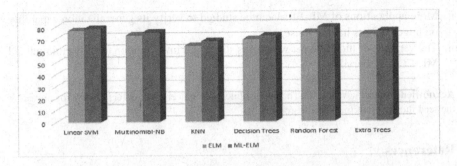

Fig. 17. F1-measure comparisons on ML-ELM and ELM Feature space (Reuters)

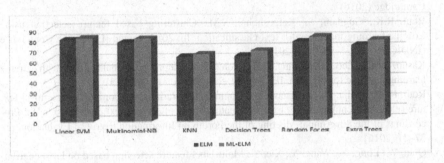

Fig. 18. F1-measure comparisons on ML-ELM and ELM Feature space (DMOZ)

5 Conclusion

The suggested approach investigates the significance of the Multi-layer ELM feature space in-depth. Initially, the corpus is subjected to a novel feature selection technique (*CORFS*), which removes superfluous features from the corpus and improves the classification performance. An extensive empirical study on several benchmark datasets has demonstrated the efficiency of the suggested technique on *HDFS-MLELM* compared to the *VS-TFIDF*. According to empirical investigations, SVM outperforms other classifiers on both feature spaces for all the datasets. After a thorough examination of the experimental results, it has been determined that the Multi-layer ELM feature space

- is able to solve the three major problems faced by the current machine/deep learning techniques as highlighted in Sect. 1.
- can replace the costly kernel techniques.
- is more suitable and much useful for text classification in comparison with *TF-IDF* vector space.

This work can be extended on the following lines:

i. Deep learning methods such as CNN, RNN, and ANN need a vast amount of data and many tuned parameters to train the network. As part of future work, combining these deep learning architectures with ML-ELM can reduce the requirement of tuned parameters without compromising their performances.

ii. More applications of ML-ELM can be studied to verify its generalization capability on huge datasets having noise.
iii. The variance of hidden layer weights is still under investigation to fully comprehend ML-ELM's operation.

Acknowledgement. : We thank Thapar Institute of Engineering and Technology for providing the seed money grant for this research work.

References

1. Goodfellow, I., Bengio, Y., Courville, A., Bengio, Y.: Deep Learning, vol. 1. MIT press, Cambridge (2016)
2. Roul, R.K., Gugnani, S., Kalpeshbhai, S.M.: Clustering based feature selection using extreme learning machines for text classification. In: 2015 Annual IEEE India Conference (INDICON), pp. 1–6. IEEE (2015)
3. Kasun, L.L.C., Zhou, H., Huang, G.-B., Vong, C.M.: Representational learning with extreme learning machine for big data. IEEE Intell. Syst. **28**(6), 31–34 (2013)
4. Roul, R.K., Bhalla, A., Srivastava, A.: Commonality-rarity score computation: a novel feature selection technique using extended feature space of elm for text classification. In: Proceedings of the 8th Annual Meeting of the Forum on Information Retrieval Evaluation, pp. 37–41 (2016)
5. Qian, W., Long, X., Wang, Y., Xie, Y.: Multi-label feature selection based on label distribution and feature complementarity. Appl. Soft Comput. **90**, 106167 (2020)
6. Zhang, L., Zhou, W.D., Jiao, L.C.: Kernel clustering algorithm. Chin. J. Comput. **6**, 004 (2002)
7. Roul, R.K., Arora, K.: A modified cosine-similarity based log kernel for support vector machines in the domain of text classification. In: Proceedings of the 14th International Conference on Natural Language Processing (ICON-2017), pp. 338–347 (2017)
8. Filippone, M., Camastra, F., Masulli, F., Rovetta, S.: A survey of kernel and spectral methods for clustering. Pattern Recogn. **41**(1), 176–190 (2008)
9. Huang, G.-B., Ding, X., Zhou, H.: Optimization method based extreme learning machine for classification. Neurocomputing **74**(1), 155–163 (2010)
10. Huang, G.-B., Chen, L.: Enhanced random search based incremental extreme learning machine. Neurocomputing **71**(16), 3460–3468 (2008)
11. Roul, R.K.: Impact of multilayer elm feature mapping technique on supervised and semi-supervised learning algorithms. Soft Comput. **26**(1), 423–437 (2022)
12. Huang, G.-B., Zhou, H., Ding, X., Zhang, R.: Extreme learning machine for regression and multiclass classification. IEEE Trans. Syst. Man Cybern. Part B (Cybern.) **42**(2), 513–529 (2012)
13. Roul, R.K., Asthana, S.R., Kumar, G.: Study on suitability and importance of multilayer extreme learning machine for classification of text data. Soft Comput. **21**(15), 4239–4256 (2017)
14. Roul, R.K.: Detecting spam web pages using multilayer extreme learning machine. Int. J. Big Data Intell. **5**(1/2), 49–61 (2018)
15. Roul, R.K.: Suitability and importance of deep learning feature space in the domain of text categorisation. Int. J. Comput. Intell. Stud. **8**(1–2), 73–102 (2019)
16. Rifkin, R., Yeo, G., Poggio, T.: Regularized least-squares classification. Nato Sci. Ser. Sub Ser. III Comput. Syst. Sci. **190**, 131–154 (2003)

17. Hartigan, J.A., Wong, M.A.: Algorithm as 136: a k-means clustering algorithm. J. Roy. Stat. Soc. Ser. C (Appl. Stat.) **28**(1), 100–108 (1979)
18. Dreiseitl, S., Ohno-Machado, L.: Logistic regression and artificial neural network classification models: a methodology review. J. Biomed. Inf. **35**(5–6), 352–359 (2002)
19. Huang, G.-B., Chen, Y.-Q., Babri, H.A.: Classification ability of single hidden layer feedforward neural networks. IEEE Trans. Neural Netw. **11**(3), 799–801 (2000)
20. Roul, R.K., Rai, P.: A new feature selection technique combined with ELM feature space for text classification. In: Proceedings of the 13th International Conference on Natural Language Processing, pp. 285–292. ACL (2016)
21. Huang, G.-B., Zhou, H., Ding, X., Zhang, R.: Extreme learning machine for regression and multiclass classification. IEEE Trans. Syst. Man Cybern. Part B (Cybern.) **42**(2), 513–529 (2011)
22. Roul, R.K., Agarwal, A.: Feature space of deep learning and its importance: comparison of clustering techniques on the extended space of ml-elm. In: Proceedings of the 9th Annual Meeting of the Forum for Information Retrieval Evaluation, pp. 25–28 (2017)
23. Roul, R.K.: Deep learning in the domain of near-duplicate document detection. In: Madria, S., Fournier-Viger, P., Chaudhary, S., Reddy, P.K. (eds.) BDA 2019. LNCS, vol. 11932, pp. 439–459. Springer, Cham (2019). https://doi.org/10.1007/978-3-030-37188-3_25
24. Roul, R.K.: Study and understanding the significance of multilayer-ELM feature space. In: Bellatreche, L., Goyal, V., Fujita, H., Mondal, A., Reddy, P.K. (eds.) BDA 2020. LNCS, vol. 12581, pp. 28–48. Springer, Cham (2020). https://doi.org/10.1007/978-3-030-66665-1_3
25. Huang, G.-B., Chen, L.: Convex incremental extreme learning machine. Neurocomputing **70**(16), 3056–3062 (2007)
26. Wang, H., He, C., Li, Z.: A new ensemble feature selection approach based on genetic algorithm. Soft Comput. **24**(20), 15811–15820 (2020)
27. Khoder, A., Dornaika, F.: Ensemble learning via feature selection and multiple transformed subsets: application to image classification. Appl. Soft Comput. **113**, 108006 (2021)
28. De Stefano, C., Fontanella, F., Marrocco, C., Di Freca, A.S.: A ga-based feature selection approach with an application to handwritten character recognition. Pattern Recogn. Lett. **35**, 130–141 (2014)
29. Tubishat, M., Abushariah, M.A., Idris, N., Aljarah, I.: Improved whale optimization algorithm for feature selection in arabic sentiment analysis. Appl. Intell. **49**(5), 1688–1707 (2019)
30. Jiang, Z., Zhang, Y., Wang, J.: A multi-surrogate-assisted dual-layer ensemble feature selection algorithm. Appl. Soft Comput. **110**, 107625 (2021)
31. Got, A., Moussaoui, A., Zouache, D.: Hybrid filter-wrapper feature selection using whale optimization algorithm: a multi-objective approach. Expert Syst. Appl. **183**, 115312 (2021)
32. Miao, J., Ping, Y., Chen, Z., Jin, X.-B., Li, P., Niu, L.: Unsupervised feature selection by non-convex regularized self-representation. Expert Syst. Appl. **173**, 114643 (2021)
33. Ezenkwu, C.P., Akpan, U.I., Stephen, B.U.-A.: A class-specific metaheuristic technique for explainable relevant feature selection. Mach. Learn. Appl. **6**, 100142 (2021)
34. Adamu, A., Abdullahi, M., Junaidu, S.B., Hassan, I.H.: An hybrid particle swarm optimization with crow search algorithm for feature selection. Mach. Learn. Appl. **6**, 100108 (2021)
35. Bengio, Y., LeCun, Y., et al.: Scaling learning algorithms towards AI. Large-Scale Kernel Mach. **34**(5), 1–41 (2007)
36. Vapnik, V.N.: An overview of statistical learning theory. IEEE Trans. Neural Netw. **10**(5), 988–999 (1999)
37. Lu, G., Gan, J., Yin, J., Luo, Z., Li, B., Zhao, X.: Multi-task learning using a hybrid representation for text classification. Neural Comput. Appl. **32**(11), 6467–6480 (2020)
38. Huan, J., Sk, A.A., Quek, C., Prasad, D.: Emotionally charged text classification with deep learning and sentiment semantic. Neural Comput. Appl. (2021)

39. Chantar, H., Mafarja, M., Alsawalqah, H., Heidari, A.A., Aljarah, I., Faris, H.: Feature selection using binary grey wolf optimizer with elite-based crossover for arabic text classification. Neural Comput. Appl. **32**(16), 12201–12220 (2020)
40. Behera, B., Kumaravelan, G.: Text document classification using fuzzy rough set based on robust nearest neighbor (frs-rnn). Soft Comput. **25**(15), 9915–9923 (2021)
41. Janani, R., Vijayarani, S.: Automatic text classification using machine learning and optimization algorithms. Soft Comput. **25**(2), 1129–1145 (2021)
42. Cheng, Y., et al.: Hsan-capsule: a novel text classification model. Neurocomputing (2021)
43. Li, X., Cui, M., Li, J., Bai, R., Lu, Z., Aickelin, U.: A hybrid medical text classification framework: integrating attentive rule construction and neural network. Neurocomputing **443**, 345–355 (2021)
44. Xu, S., Xiang, Y.: Frog-gnn: multi-perspective aggregation based graph neural network for few-shot text classification. Expert Syst. Appl. **176**, 114795 (2021)
45. Tang, Z., Li, W., Li, Y.: An improved supervised term weighting scheme for text representation and classification. Expert Syst. Appl. **189**, 115985 (2022)
46. Mou, S., Du, P., Cheng, Z.: A brain-inspired information processing algorithm and its application in text classification. Expert Syst. Appl. **177**, 114828 (2021)

ARCORE: A Requirements Dataset for Service Identification

Vijaya Peketi[✉] and Surekha Satti

Hyderbad, India
vijaya.peketi@gmail.com

Abstract. Service Oriented Software Engineering (SOSE) has been an integral process for developing medium to large scale software. However, identifying the needed services is largely dependent on expertise of software architects. We are interested in assisting architects by automating the process of service identification from requirements documents. One of the mechanisms to automate the service identification process is using Machine Learning Techniquesand it requires working with large datasets. In this paper, we detail our work on creating a requirements dataset called ARCORE, that can be used for service identification. The motivation arises from the fact that no such dataset is available in public domain, and very minimal work has been done on automated service identification from requirements documents.

Keywords: Services based software engineering · Component based software engineering · Requirements · Service · Software design · Software architecture · Dataset · Multiclass labeling · Machine learning · Artificial intelligence · Natural language processing

1 Introduction

Modern day software systems tend to use services that are either mined, developed internally or contracted/bought from third parties. A service is defined as a unit of functionality that can run independently [1]. A service may use one or more components that depend on each other. Alternatively a component may call one or more services. A software component can be defined as a unit of software that encapsulates certain functionality. Component Based Software Engineering (CBSE) approach, that gained popularity in the 90's, focuses on software reuse and suggests building applications with reusable components. In our work, we view components as mechanisms of realization of services and shall be using the terms interchangeably.

Munialo et al. [2] argue that upto 40% of effort can be saved if the component needed for implementation is readily available. While work exists on appropriateness of the chosen component within the application, component testing, and degree of reusability of components [3], there isn't much work on identifying the components from the requirements. One of the reasons could possibly be the lack

---

V. Peketi and S. Satti—Independant Consultant.

P. P. Roy et al. (Eds.): BDA 2022 India, LNCS 13773, pp. 53–67, 2022.
https://doi.org/10.1007/978-3-031-24094-2_4

of suitable requirements dataset, and approaches that automate identification of components. In this paper, we discuss the creation of one such software requirements dataset called ARCORE[1]. There are two popular software requirements datasets available in literature. They are PROMISE [4] and PURE [5] datasets. However these datasets are not tailored towards CBSE and aren't suitable for identification of components. We have created a software requirements dataset[2] that can be used for identification of component categories.

ARCORE dataset has been created from 11 publicly available Software Requirements Specification (SRS) documents 3 retrieved from Web. The documents retrieved are created during last 10 years to ensure that these documents have sufficient Service Cues for identification of Component categories. A hint of software component category in the Requirements text is referred as an Service Cue. A component may be bucketed under a certain category based on the type of functionality accomplished. Bucketing it under a certain category helps maintain the component registry and can eventually help with component discovery. For example, Compute, Storage, Integration, Control, User Interface are some possible component categories. If we consider a requirements statement like, "The static files are stored as needed", the word "stored" is a possible Service Cue for component category *Storage*. If we consider the requirements statement, "Approval role has view privileges", the words/phrases "approval", "view privileges" are possible Service Cues for *Security*.

As part of creation of the ARCORE dataset, we have identified a set of Service Cues. Then we got validated them with experienced architects. The validated Service Cues along with instructions of usage and sample scenarios have been included in Annotation Guide. Using the Annotation guide data Annotation was performed with 120 annotators. An extension of this work is to create AI solution which can identify Component Categories for Requirement statements. We believe the dataset has other application usage prospects too. For example, the dataset can be used for information discovery from requirements. It can be used for functional requirements (FR) and non-functional requirements (NFR) classification, detailed study on any FR or NFR Classification, etc.

The remaining paper is structured as follows: Sect. 2 gives an overview of the related work on requirements datasets, identification of component categories and classification of requirements. Section 3 details our approach of preparing ARCORE Dataset. Section 4 discusses the limitations and some possible future work.

2   Related Work

2.1   Requirements Datasets:

The International Software Benchmarking Standards Group (ISBSG)[3] Repository and PRedictOr Models In Software Engineering (PROMISE) [6], PUblic

---

[1] https://tinyurl.com/ARCOREDataset.
[2] Paper is offered for single blind review process.
[3] https://www.isbsg.org/isbsg-academic-research-information.

Requirements Dataset (PURE) [5] and RALIC [7] are some of the datasets used by the Software Engineering community. PROMISE dataset is a public repository inspired by the University of California Irvine Machine Learning Repository, designed to promote repeatable, verifiable, disputable, and/or improved software engineering prediction models [9]. PROMISE has been most frequently (11/24 or 45%) used by researchers [10]. The original repository consists of a pre-labeled set of 255 FRs and 370 NFRs, the last one being sub-classified into 11 different types of NFRs. PROMISE dataset was created 10 years back by Graduate students of DePaul university as part of term project work.

PURE dataset, an extension of PROMISE, is suitable for model synthesis, abstraction identification and document structure assessment. Soo Ling Loom created RALIC, a requirements dataset from different projects from Univeristy College of London as part of her thesis [7]. According to Binkhonain et al [10], there are very few shared/publicly available datasets and, these are provided by academia. Also a researcher interested in using the datasets must perform his own research about the sources from which the dataset was created. Hence there is a need for datasets that focus on SRS documents from industry. Additionally, PROMISE, PURE and RALIC are not suitable for Service Categorization work.

2.2 Service Selection

In the development of service-oriented architecture-based applications (SOABA), service selection is a crucial but extremely difficult process [33]. The user's business needs and available money are considered while choosing a service. Several service providers put a lot of effort into choosing the most popular service from a pool of comparable services offered on the market.

2.3 Requirements Classifications

Requirements Classification is an area of interest to Researchers and Academia to handle various tasks in Requirements Engineering (RE) [29]. There have been studies done on a variety of topics in RE, including the classification of functional and non-functional requirements, traceability, the detection of equivalent requirements, ambiguity detection, and model synthesis. Primarily the requirements are classified as Functional and Non-Functional requirements [NFR] [18]. Researchers have applied Natural Language Processing (NLP) and Machine learning techniques to automate the classification of Requirements [11, 15]. Zayed in his literature study found that there are very few studies based on Automatic classification of Software Requirements [16]. The ARCORE dataset can potentially be used as training dataset to automate the classification of other SRS documents.

2.4 Techniques Used for Automatic Requirements Classification

Canedo et. al [8]tested various combinations to compare the accuracy, recall, and F-measure of the categorization findings. With an F-measure of 91% on the binary classification, 74% in the 11 granularity classification, and 78% on the 12 granularity classification, we discovered that the combination of TF-IDF and LRhas the best performance measures for binary classification, non-functional requirements classification, and for requirements classifications in general. In order to classify the software requirement (SR) specification, the binary classification of SRs into functional requirement (FRs) was performed by Rhimi et. al, [31] study using three ensemble approaches: accuracy as a weight ensemble, mean ensemble, and accuracy per class as a weight ensemble with a combination of four different DL models—long short-term memory (LSTM), bidirectional long short-term memory (BiLSTM), a gated recurrent unit (GRU), and a convolutional neural network (CNN). The PROMISE dataset was used to train and test the models. To categorise SRs into one of the 12 multi-classes of FRs and NFRs, a one-phase classification system was created. A two-phase classification scheme was also created to categorise SRs.

3   ARCORE Dataset

The ARCORE Dataset has been created using Content Analysis Method as suggested by Neuendorf et. al [32]. Content Analysis Method is a research technique used to identify the existence of specific words, topics, or concepts in a given set of qualitative data (i.e. text). Researchers can quantify and examine the occurrence, significance, and connections of such specific words, themes, or concepts using content analysis. For instance, academics can assess the language used in a news story to look for partiality or bias. The meanings contained in the texts, the author(s), the audience, and even the culture and time period surrounding the text can all be inferred by researchers. Sources of data could be from interviews, open-ended questions, field research notes, conversations, or literally any occurrence of communicative language (such as books, essays, discussions, newspaper headlines, speeches, media, historical documents). A single study may analyze various forms of text in its analysis. To analyze the text using content analysis, the text must be coded, or broken down, into manageable code categories for analysis (i.e. "codes"). Once the text is coded into code categories, the codes can then be further categorized into "code categories" to summarize data even further.

The process followed for creating ARCORE Dataset is shown in Figure 1. In the diagram, vertical swim lanes represent activities performed by different stakeholders and the corresponding artifacts produced. For example, the first swim lane depicts actions performed by experienced Architects and the resultant artefact. Table 1 outlines the experience of the architects and their expertise. The Second Swim lane represents actions performed by authors and the third Swim lane has the activities performed by Annotaters.

There are four stages in ARCORE Dataset creation Process 1. The description of the steps in each stage is given below.

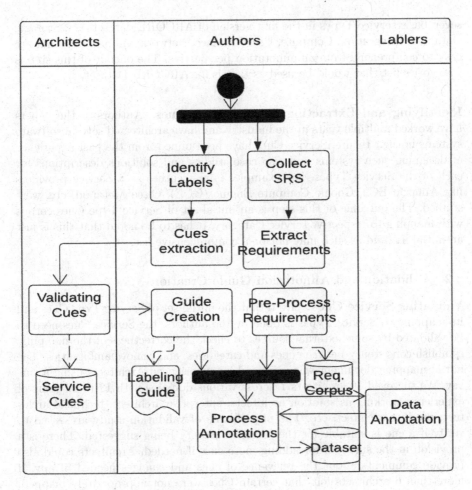

Fig. 1. Dataset creation process

3.1   Service Cues Creation

Selecting Service Types. As per the Amazon Web Serices (AWS) white paper [19], AWS comprises of about 200 products and services including computing, storage, networking, database, analytics, application services, deployment, management, machine learning, mobile, developer tools, and tools for the Internet of Things. Similarly, there are multiple services that other cloud service providers like Microsoft, Google, HP, etc. provide. Figure 2 is a list of various categories of major cloud service market place providers. The categories are provided by the vendors themselves. The most common of the categories, Storage, Database, Integration, Compute, Network and Security have been identified for classification. User Interface is also chosen as it is also commonly available category in service requirements classification. Creating an annotated dataset for all the services currently available in market is a humongous task. Hence, we started with seven basic service Types in the first version of ARCORE dataset, i.e., Storage, Database, Integration, Compute, User Interface, Network and Security, that are easy to comprehend from an annotation perspective. The output of this step is Categories list that would be used as Labels for ARCORE Dataset.

Identifying and Extracting Service Type Cues. Authors of the papers have worked multiple years in the industry and have architected several software systems in their tenure. Service Cues have been mined from the past experience of designing such systems and also case studies and solutions descriptions for each of the Service Types. For example, Infrastrcuture as a Service providers like Amazon EC2, Google Compute Engine (GCE), Cisco Metapod, etc. were studied. The outcome of this step is an initial set of Service Type Cues corpus with mapping to respective service Category. It has to be noted that this is just an initial list and as such may not be a comprehensive list.

3.2   Validation and Annotation Guide Creation

Validating Service Cues. Given that the initial set of Service Type Cues and its mapping to Service Types is done by the authors, the Service Cues need to be validated by some external sources to check the correctness of the mapping, establish consistency between cues and categories, and remove ambiguities. The initial mapping document was validated by experienced Architects using a survey. We surveyed 10 architects 1 across various domains with 12 to 22 years of experience working as solution architects, enterprise architect, technical architect, product architect, etc. The participants of validation study are asked to verify if a cue is suitable for the service Category being suggested. There is a provision in the survey 3 for adding more cues if needed. Architects could also provide comments about the relevance of cues and changes needed, if any. If more than 6 architects felt that certain Cues were not appropriately mapped, they were removed from the final list. Also, suggestions for additions of new cues were done even if one architect recommended it. The outcome of this step is a corpus of validated Service Cues and its service category mapping.

The Service Cues Corpus is validated by Architects in the mode of Survey. Based on survey output, we developed an "Service Cues" repository. The repository constitutes of mapping of "Service Cues" with respective Service Type (Table 2).

GCP	AWS	Azure
AI and Machine Learning	Analytics	AI + machine learning
API Management	Application Integration	Analytics
Compute	Blockchain	Compute
Containers	Business Applications	Containers
Data Analytics	Cloud Financial Management	Databases
Databases	Compute	Developer tools
Developer Tools	Contact Center	DevOps
Healthcare and Life Science	Containers	Hybrid + multicloud
Hybrid and Multicloud	Database	Identity
Internet of Things	Developer Tools	Integration
Management Tools	End User Computing	Internet of Things
Media and Gaming	Front-End Web & Mobile	Management and governance
Migration	Games	Media
Networking	Internet of Things	Migration
Operations	Machine Learning	Mixed reality
Security and Identity	Management & Governance	Mobile
Serverless Computing	Media Services	Networking
Storage	Migration & Transfer	Security
	Networking & Content Delivery	Storage
	Quantum Technologies	Virtual desktop infrastructure
	Robotics	Web
	Satellite	
	Security, Identity, & Compliance	
	Serverless	
	Storage	

Fig. 2. Service Categories

Creating Annotations Guide. Prior to the distribution of requirements statements forms to annotators, an annotation guide is created for documenting ground rules and rationale for performing data annotation. The Annotation guide consists of explanation for using Service Type Cues for performing annotation task.

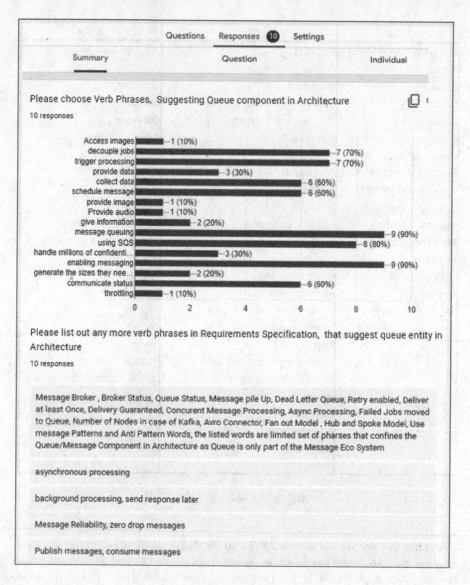

Fig. 3. Service cues survey

3.3 Requirement Corpus Creation

Gather Publicly Available SRS Documents. This process is done in parallel to Service Type Cues Creation Process as depicted in Fig. 1.

As part of identifying SRS Documents for performing annotation, authors studied Software Requirement Specification (SRS) documents available in public domain. These documents were obtained by querying web sources and digital databases. A literature review was been performed to extract the docu-

Table 1. Survey participants details

Participant ID	Experience	Architecture area
A1	20	Product, Solution
A2	15	Product
A3	18	Application, Solution
A4	12	Solution
A5	10	Application
A6	15	Solution, Enterprise
A7	19	Product, Application
A8	22	Solution, Technical
A9	16	Enterprise
A10	12	Solution

Table 2. Service cues corpus details

Category	Service cues	Validated enhanced
Database	20	30
Storage	15	12
Integration	18	24
User interface	12	16
Compute	10	15
Network	15	10
Security	19	15

ments using search strings such as "Software requirement documents", "Software requirements specification government", "SRS gov", and "SRS Federal". Additionally, we shortlisted different domains like Payments, Video streaming, Pharma etc. The intent of the search was to extract relevant documents that are representative of certain domains and thereby form a basis to create a dataset rather than perform an exhaustive systematic literature review and retrieve all publicly available requirements documents in all domains. The details of the requirements documents and the corresponding domains are provided in Table 3.

Extracting Requirement Statements. Requirement statements were extracted from the shortlisted SRS documents for creation of Requirements Corpus. From the 11 SRS documents, about 800 statements were extracted.

Refining Requirements. The input for this step is extracted Requirement statements. The statements are refined to ensure complex statements are converted to simple statements and atomicity of the statement is maintained. For example, statements connected by coordinating conjunctions have been broken

Table 3. SRS documents details

Project	Requirements	Demography	Technology	Business	Source
eDistrict	51	India	Cloud	eGovernance	it.delhi.gov.in
LTP	40	EU	Cloud	Registration	www.nlcs.gov.bt
BalticLSC	119	EU	IaaS	Computing	www.balticlsc.eu
CloudLSVA	81	USA	Streaming	Automobile	Cloud-lsva.eu
CTMS	50	Singapore	Proj. Mgmt	Construction	edata.sg
eProcu.	70	Africa	Web	Bid Mgmt.	www.adb.org
eRostering	121	UK	ERP	Healthcare	www.england.nhs.uk
GBS	51	Norway	Analytics	Statistics	www.ssb.no
NIST	60	USA	ERP	Security	www.nist.gov
UPI	53	India	API	Payments	www.npci.org
Network	109	Global	Network	Products	VPC

to two simple statements. As a result, the possibility of each statement containing multiple Service Type Cues and correspondingly multiple service category mappings is reduced.

3.4   ARCORE Dataset Creation

As a next logical step after creating requirements corpus, we created examples of the requirements statements belonging to different service Types.

Creating Forms for Annotators. We created Google Forms for capturing Annotations for each Requirement Statement in Dataset. The forms were distributed to 120 participants in such a way that at least seven participants annotate a requirements statement. This technique helps us eliminate any bias, that could be introduced as part of Annotation task. A set of more than 70 forms were created and distributed among participants. Annotation guide was provided for aiding the annotation task. The Guide had examples of complex scenarios too. Participants were also provided with links to original SRS document that has been used for specific requirements in the form. They were encouraged to study the document for understanding the context of the project.

Performing Annotations. Data annotation was performed through Crowd sourcing by a closed group of 120 members. The survey participants were graduate students in reputed academic institute, had varying amount of industrial experience and had knowledge on designing software systems. The motivation for performing crowd sourcing activity is to annotate the requirements with the respective service category labels along with rationale. The outcome of this step is multiple annotation sets provided by about 7 participants for each of the requirement statements and associated rationale for choice of annotation from each of them.

Survey Details: A total of 120 people took part in this survey, with the majority of them being Research Scholars and Masters students from reputable institutions. The majority of the participants have worked in the industry on medium and large-scale systems. It takes a lot of time and effort to annotate 800 statements. As a result, the task is broken down into smaller chunks. In the form of survey forms, the corpus of 800 statements is divided among 120 members. The rule stated that at least six people had to annotate a single requirement sentence. The most common method for obtaining good quality text span annotations is to collect numerous redundant annotations, which are then combined into a higher-quality meta-annotation [21].

We described the aim of the survey, the survey process, and how to use the labelling user guide to annotate data before performing the survey. Details of the SRS document were included in the survey forms to help people grasp the context and deeper significance of the software project. Before filling out the survey form, survey participants must read the SRS document. Each required sentence requires the participant to choose the labels and explain why they were chosen. Table 4 is a sample of survey result provided by participants for a single Requirement Sentence. About 70 survey forms have been created for capturing Labels for each of items in Requirement Corpus. Two software developers classified a representative samples to see whether or not this was a big issue. Utilizing Randolph's Online Kappa Calculator [22], we had a free-marginal kappa of 0.72, indicating adequate inter-rater agreement, by comparing the first authors selections against the majority vote of the other annotators. Majority voting scheme was used to resolve any sort of disagreements.

3.5   Sample Response Explanation

Requirement Text: "On selecting a particular application form the dealing officer can see the details of the application form along with the attached documents."

Participant 1 Labels: User Interface; Storage;
Participant 1 Rationale: Select keyword in the context is related to User Interface and attached documents are concerned with Storage

Participant2 Labels: Database; Storage;User Interface; Compute
Participant2 Rationale: Fetching out application details is associated with the database. Fetching documents attached is associated with Storage. It is a microservice so compute is associated.

Processing Annotated Requirements. Given each requirements statement was annotated by multiple participants, majority voting technique was used for finalizing the annotations for each statement in Dataset, i.e. at least four annotators giving a judgment on any requirements is considered. Survey results are analysed and verified as part of this task. Table 5 has details of number of annotations per category.

Table 4. Responses for a survey question

Req. text	On selecting a particular application form the dealing officer can see the details of the application form along with the attached documents
Labels	Rationale
UI;	hints user interfaces
DB; STRG; INTG; UI;	Cross reference Select keyword is related to User Interface and attached documents are concerned with Storage
DB; CMPT; STR; UI;	Fetching out application details is associated with database. Fetching documents attached is associated with Storage. It is a microservice so compute is associated
UI; DB;	User interface for viewing information, database for retriving details
STRG; UI; INTG;	Storage required for the forms and interface is needed to view them. Integration of the systems is a must
UI; STRG; DB;	Display of content using user interface to display contents of database query
STRG; DB; INTG;	The details will be fetched from database and documents will be fetched from storage
DB; UI;	Fetching details (Database), Seeing details (UI)
INTG; UI;	The requirement to display falls under user interface and the requirement to show the attached documents needs integration of user interface and backend storage
UI ;	See the details of the application form
Legend	UI::UserInterface; DB::Database; STRG::storage; CMPT::Compute; INTG::Integration;

Table 5. Annotation data distribution

Serial Number	Data Annotation Name	Requirement Count
1	Database	343
2	Storage	381
3	Integration	372
4	User Interface	369
5	Compute	258
6	Network	169
7	Security	262
8	Others	17

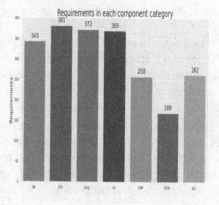

4   Conclusion and Future Work

ARCORE Dataset creation is a step towards automating identification of Service Specific requirements from software requirements. During our study of Software Requirements specifications we have discovered the existence of FR and NFR categories which can also be used for classification F, NFR and NFR subclassifications.

The ARCORE dataset inherits general limitations of annotated datasets such as personal bias and knowledge limitation of the participants for specific contexts of each requirements statement. The requirements statements in the dataset may have correctness and completeness concerns which, in turn, might confuse annotators. Besides these limitations, the dataset is useful for multiple applications as indicated in Sect. 1.

ARCORE dataset can be used in creating NLP and Machine learning solutions in SE. We can create ML models using this dataset. The models can be trained, tested and validated for automated identification of Service Types from requirements. Unsupervised machine learning algorithms can be applied to ARCORE Dataset for training, validation and we can apply Deep Learning algorithms to overcome the poor generalization problem in classification tasks.

We intend to enhance the Dataset with more categories in future to cover various products available in the market for ready use. We also plan to make the labels more fine grained considering categories within DB such as SQL, NoSQL, etc. The original intent is to expand for all Service Types available in marketplace. However, there are possible use-cases of Dataset in FR and NFR classifications too.

Acknowledgements. This paper and the research behind it would not have been possible without the exceptional support of our supervisor, Dr. Raghu Reddy, Head of SERC, IIIT Hyderabad. His enthusiasm, knowledge and exacting attention to detail have been an inspiration and kept our work on track to the final draft of this paper.

Note from Authors: Evidences and intermediate Artefacts created for each of the steps shall be made available to reviewers on need basis.

References

1. Sommerville, I.: 10th edition of Ian Sommerville's Software Engineering, chapter 18. Pearson (2016)
2. Muketha, G., Omieno, K.: An effort estimation method for service-oriented architecture. J. Eng. Sci. Technol. Rev. **13**, 186–196 (2020). https://doi.org/10.25103/jestr.136.25
3. Bengtsson, A., Nielsen, P., Li, M.: Component trustworthiness in an enterprise software platform ecosystem. In: Conference: NOKOBIT (Norsk IKT-konferanse for forskning og utdanning) At: NTNU, Trondheim, Norway (2021)
4. Lima, M., Valle, V., Costa, E., Lira, F., Gadelha, B.: Software engineering repositories: expanding the PROMISE database. In: SBES 2019: Proceedings of the XXXIII Brazilian Symposium on Software Engineering, pp. 427–436 (2019). https://doi.org/10.1145/3350768.3350776

5. Ferrari, A., Spagnolo, G., Gnesi, S.: PURE: a dataset of public requirements documents. In: Conference: 2017 IEEE 25th International Requirements Engineering Conference (RE), pp. 502–505 (2017). https://doi.org/10.1109/RE.2017.29
6. Cheikhi, L., Abran, A.: Promise and ISBSG software engineering data repositories: a survey. In: Conference: Software Measurement and the 2013 Eighth International Conference on Software Process and Product Measurement (IWSM-MENSURA), 2013 Joint Conference of the 23rd International Workshop on Project: Software Estimation Authors, pp. 17–24 (2013). https://doi.org/10.1109/IWSM-Mensura. 2013.13
7. Lim, S.L.: Social networks and collaborative filtering for large-scale requirements elicitation. Disseration University of New South Wales (2011)
8. Canedo, E.D., Mendes, B.C.: Software requirements classification using machine learning algorithms. Entropy **22**(9), 1057 (2020)
9. Shirabad, J.S., Menzies, T.: The PROMISE Repository of Software Engineering Databases (2005)
10. Binkhonain, M., Zhao, L.: A review of machine learning algorithms for identification and classification of non-functional requirements. Expert Syst. Appl. **1**, 100001 (2019)
11. Sharma, R.: Grounding functional requirements classification in organizational semiotics. In: Interdisciplinary Approaches to Semiotics, vol. 151 (2017)
12. Zhang, Y., Harman, M.: Search based optimization of requirements interaction management. In: 2nd International Symposium on Search Based Software Engineering, pp. 47–56. IEEE (2010)
13. Gholamshahi, S., Hasheminejad, S.M.H.: Software component identification and selection: a research review. Softw. Pract. Exp. **49**(1), 40–69 (2019)
14. Márquez, G., Astudillo, H.: Selection of software components from business objectives scenarios through architectural tactics. In: 2017 IEEE/ACM 39th International Conference on Software Engineering Companion (ICSE-C), pp. 441–444. IEEE (2017)
15. Iqbal, T., Elahidoost, P., Lucio, L.: A bird's eye view on requirements engineering and machine learning. In: 2018 25th Asia-Pacific Software Engineering Conference (APSEC), pp. 11–20. IEEE (2018)
16. Zayed, M.A.: Automatic software requirements classification: a systematic literature review (2021)
17. Heidari, A., Navimipour, N.J.: Service discovery mechanisms in cloud computing: a comprehensive and systematic literature review. Kybern. J. (2021)
18. Sharma, V.S., Ramnani, R.R., Sengupta, S.: A framework for identifying and analyzing non-functional requirements from text. In: Proceedings of the 4th International Workshop on Twin Peaks of Requirements and Architecture, pp. 1–8 (2014)
19. Mathew, S., Varia, J.: Overview of Amazon Web Services-AWS Whitepaper (2021)
20. Liu, J., Chang, W.C., Wu, Y., Yang, Y.: Deep learning for extreme multi-label text classification. In: Proceedings of the 40th International ACM SIGIR Conference on Research and Development in Information Retrieval, pp. 115–124 (2017)
21. Zlabinger, M., Sabou, M., Hofstätter, S., Hanbury, A.: Effective crowd-annotation of participants, interventions, and outcomes in the text of clinical trial reports. Findings (2020)
22. Randolph, J.J.: Online Kappa Calculator (2008). https://justusrandolph.net/kappa/. Accessed 02 Sept 2022
23. Sabou, M., Bontcheva, K., Derczynski, L., Scharl, A.: Corpus annotation through crowdsourcing: towards best practice guidelines. LREC (2014)

24. Zlabinger, M., Sabou, M., Hofstätter, S., Hanbury, A.: Effective crowd-annotation of participants, interventions, and outcomes in the text of clinical trial reports. In: Findings of the Association for Computational Linguistics: EMNLP 2020, pp. 3064–3074 (2020)

25. Serban, A., van der Blom, K., Hoos, H., Visser, J.: Adoption and effects of software engineering best practices in machine learning. In: Proceedings of the 14th ACM/IEEE International Symposium on Empirical Software Engineering and Measurement (ESEM), pp. 1–12 (2020)

26. Vathsavayi, S., Räihä, O., Koskimies, K.: Tool support for software architecture design with genetic algorithms. In: 2010 Fifth International Conference on Software Engineering Advances, pp. 359–366. IEEE (2010)

27. Ali, A., Shamsuddin, S.M., Eassa, F.E.: Ontology-based cloud services representation. Res. J. Appl. Sci. Eng. Technol. 8(1), 83–94 (2014)

28. Heidari, A., Navimipour, N.J.: Service discovery mechanisms in cloud computing: a comprehensive and systematic literature review. Kybernetes (2021)

29. Ferrari, A., Spagnolo, G.O., Gnesi, S.: Towards a dataset for natural language requirements processing. In: REFSQ Workshops (2017)

30. Dias Canedo, E., Cordeiro Mendes, B.: Software requirements classification using machine learning algorithms. Entropy 22(9), 1057 (2020)

31. Rahimi, N., Eassa, F., Elrefaei, L.: One-and two-phase software requirement classification using ensemble deep learning. Entropy 23(10), 1264 (2021)

32. Neuendorf, K.A.: The Content Analysis Guidebook. Sage, Thousands oaks (2017). https://doi.org/10.4135/9781071802878

33. Siddiqui, Z.A., Tyagi, K.: Study on service selection effort estimation in service oriented architecture-based applications powered by information entropy weight fuzzy comprehensive evaluation model. IET Softw. 12(2), 76–84 (2018)

Learning Enhancement Using Question-Answer Generation for e-Book Using Contrastive Fine-Tuned T5

Shobhan Kumar[✉][iD], Arun Chauhan[iD], and Pavan Kumar C.[iD]

Indian Institute of Information Technology Dharwad, Dharwad, Karnataka, India
{shobhank9,arun.chauhan,pavan}@iiitdwd.ac.in

Abstract. The rise of E-learning systems offers a vast amount of free educational content for every inquisitive e-learner. However, reading only e-content does not makes their learning effective. Posing appropriate questions during the reading process can aid in the learner's comprehension. We present a novel approach to create educational Question-Answers on eBook content. The model first summarizes key information from an input document, which is then used for creating relevant Question-Answers (QAs). We build our educational Question-Answers generation model by fine-tuning a pretrained LM (T5) using maximum likelihood estimation. We also present a contrastive fine-tuning method, in which the contrastive loss is added between the positive and negative training feature pairs during the fine-tuning process. The extensive evaluation methods on QA dataset (FairytaleQA and HotPotQA) and NCERT e-book, demonstrate the effectiveness and practicability of our eQA model.

Keywords: Question-Answers (QA) · Question Generation (QG) · Text-summarizer · Contrastive leaning · e-Learning

1 Introduction

Textbooks is an primary source of information for students. Besides this students tends to study e-books, lecture notes, and MOOCs for further knowledge acquisition. With these passive reading materials students can only partially understand the material presented, and it does not makes their learning effective [12], it may have a major negative effect on their academic careers. In order to maximise students long-term learning trajectories, it is essential to actively engage students with their reading material. Asking questions about the reading content and evaluate the answer is a intuitive way of promoting learning [17,20,47]. Even adults find it difficult to pose educationally significant questions to engage children in reading e-books, since they lack the knowledge or time to do so [14]. This motivates us to look into ways to generate educational question-answers for e-book content to aid the students learning.

The proposed educational Question-Answer Generation (eQAG) model presents a set of Question Answers (QAs) for each chapter of the e-book. Teachers might use these QAs for self-study, before discussing the subject in class which would helps them to focus their in-class discussion for deep knowledge transfer to the students. The development of automated methods to help teachers effectively create assessment questions

P. P. Roy et al. (Eds.): BDA 2022 India, LNCS 13773, pp. 68–87, 2022.
https://doi.org/10.1007/978-3-031-24094-2_5

for students is one key real-world use of question- generator [21,47]. In this work, we present a QA generation framework, that combines abstractive summarizing and QA-generation for e-book contents, that makes the students learning more effective. We assess our methodology on QA dataset (HotPot QA [49], FairytaleQA [48]) as well as PRML [2] and NCERT eBooks[1] The learner can use our QA generation model for self-assessment questions, QAs to gauge their conceptual understanding.

Fourfold contributions of this work are as follows:

- Text summary generation: We train the t5 transformer to extract the informative sentences that are most likely for educators to design question-answers for the original input.
- Contrastive training of t5_eQG: We train the t5 transformer on FairytaleQA, HoTpoTQA dataset. Contrastive loss is added between the positive and negative training feature pairs during the fine-tuning process. It helps in generating more complex questions on input document.
- Question-Answer generation (eQAG) task: We design an educational QA generation (eQAG) model for a eBook at the various level and grade. The model generate the answer for all the questions of QG task. These QA pairs can be used to evaluate a student's comprehension ability.
- Discriminator for relevance check: We trained the BERT discriminator using the ground-truth QAs and the generated negative samples. The discriminator determines whether or not the generated question, QAs is pertinent to the given passage. This ranking module(discriminator), is used to rank and choose the best potential questions and QA-pairs for the given text.

The majority of earlier research on Question Generation (QG) concentrates on producing questions based on predetermined answer spans or "keywords" [5,19,34]. It can produce low cognitive level questions that are factual in nature and are based on the local context. These models struggle to handle high cognitive level questions where we must identify the key events and comprehend the relationships among various elements/events. We used the FairytaleQA data set to train the our QAG model. Each question belongs to one of several types, some of which require significant cognitive effort, such as "activity" and "causal relationship". This enables the generation of more challenging educational QAs and makes the students e-book reading more effective.

2   Background

In recent years, the natural language processing communities have shown a great deal of interest in the Question Generation (QG) task, which creates a natural question corresponding to the input text or answer phase.

Previous research has explored the importance of QA model in teaching learning process [13,22,23]. In order to automatically generate questions from textbooks, Wang et al. [45] proposed QG-Net model that uses a pointer-generator model trained on SQuAD. A structured ontology has been used to generate educational questions [41].

---

[1] https://ncert.nic.in/textbook.php.

QuizBot model proposed by Ruan et al. [38] offers an dialogue interface to exposes set of generated questions to the students. The syntactic cues have been used in the rule based model to create queries [6]. Declarative sentences were transformed into natural questions by the emergence of sequence-to-sequence models such as Du et al. [11] and Radford et al. [37]. Compared to earlier approaches, they were more successful. Applying pre-trained transformers [4] and various optimization objectives [35] led to further advancements.

The QG task involves gathering important instructional data (Content Selection) and turning them into questions. In recent years, statistical and neural network-based algorithms have been used to identify significant passages and concepts [4,11] and produce questions regarding their selection [10]. In more recent work, a backtranslation tool was paired with a syntactic question generator to eliminate grammatical errors and increase robustness [9]. Syn-QG can consistently generate different forms of causal questions using two distinct patterns. But the syntactic restrictions hamper the system's ability to generate diverse questions. Large transformer models [3,35] and specialised attention mechanisms [51] are two methods of question generation. It has also been investigated to train QA and QG models simultaneously [44].

A few studies are concentrating on developing questions based on the information fusion of many sentences or documents - Liangming et al. [44], Yuxi et al. [31] and Luu et al. [42]. For QA generation, NarrativeQA proposed by Kocisky et al. [18] aims to incorporate important information from many places inside a paragraph. Similar to this, the MS MARCO proposed by Nguyen et al. [30] dataset combines many sources of responses to search queries. The employment of a reinforcement learning agent to align questions from various documents is proposed as a contrastive strategy by Woon Sang et al. [5], where supervised model is trained to produce questions on a text. To achieve good performance, questions with summaries and reports were generated using a rule-based methodology by Chenyang et al. [29]. The solutions discussed above typically don't take the educational component into account and could not be effective in the real world edu QG task. Our research focuses on the generating QAs on e-book content, so in this work we used FairytaleQA dataset [48]. For each paragraph in FairytaleQA, experts typically create a different style of question. We propose that context is a key factor in determining the kinds of questions that ought to be made while reading e-books.

A document's fine-grained hierarchical relations, such as actions and causal links, cannot be captured by abstractive approaches based on encoder-decoder architectures since they typically encode an input text token-by-token sequentially [39]. As a result of their capacity to represent the intricate relationships seen in a document, Graph Neural Network (GNN) models [26], are now being used in summarization research. When extracting sentences from a text, the discourse-level dependency graph that was used to encode it is decoded [46]. Similar to this, token-level and sentence-level relations in a document are both encoded using the heterogeneous graph, which is then used to extract sentences [43]. Summarizing key events from a paragraph to create instructive questions is still a challenge in the field of education.

The generated QA from rule based QA model suffered from the lack of variation [24]. In recent years, the neural-based models are used for QG tasks [10,40]. However these model have not focused on educational context.

The main motive of this work is to create a QAG system that will produce high-quality questions, QA-pairs so that the a parent or teacher could pose to children while tutoring them. Our model is trained on FairytaleQA [48] and HotPotQA dataset. The FairytaleQA dataset has 10,000 QA-pairs from 278 text document, focuses on narrative comprehension for elementary to middle school students.

We propose three-step pipeline for eQAG model: To begin, we trained the T5 summarizer to extract the important details from the provided passages of the eBook. Next, we trained the T5_QAG model to create pertinent questions and answers that correspond to the generated summary. Finally, use the BERT discriminator to rank the top QA-pairs for each text document using a predetermined threshold.

3   Methodology

The Fig. 1 depicts the overall architecture of our educational QA generating system for e-books, which consists of two modules: educational QA generation and abstract summary creation. We first create summaries of type s with the input paragraph d, and then generate the QAs q with the associated summaries. The generated QAs are said to be relevant if the question q_i can be answered with the paragraph d_i and this is considered as maximizing the conditional probability $p(q|d)$:

$$q = \mathrm{argmax}(p(q|d)) = \mathrm{argmax} \prod_{i=1}^{L} p(w_i|d, q_{i'}) \tag{1}$$

where w_i is the ith token at time stamp t of the question q, and $q_{i'}$ denotes the prior tokens (at $t-1$ time stamp), i.e., $q_1, \ldots, q_{i-1}$. As it was mentioned in the introduction, the exposure bias issue affects QA generator. Therefore, to increase the relevancy score of the generated questions, we introduced a contrastive loss and fine tune the T5 model utilising both positive and negative instances of the training data.

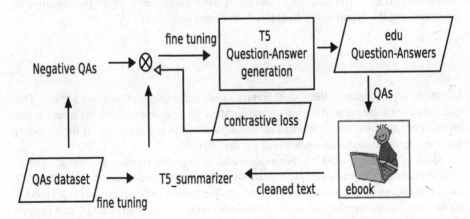

Fig. 1. The abstractive summarizer and question-answer generating module make up the overall architecture of our educational-QAG model.

3.1 T5 - Abstractive Summarizer

In this work, we examined the summary generation, and question-answer generation as a task of text-to-text transformation. As a result, we first train a T5 (Text-to-Text Transformer) summarising model to produce the abstract summary of the input text [19]. The t5 model takes text as input and produces an abstract summary of input text.

In the FairytaleQA dataset, a single paragraph typically contains many questions of various categories, and answer for a single question may be dispersed over several portions. As previously stated, we believe that context greatly influences the development of complex questions, and that high-level cognitive questions revolve on significant events and relationships.

It's challenging to find the golden summary to train the abstract summarizer. FairytaleQA, a question-answer (QA) dataset, offers both the questions and the answers. We employed the rule-based approach [7] to annotate the QA samples and then utilised it as silver summaries to train our summarization model. The rule-based approach, substitutes responses for question words in the semantically processed questions and ignores the keywords of the question.

3.2 Edu Question-Answer Generation (eQAG)

Once the model produce the abstract summary of the input text, next step is to generate an educational question, QA out of it. We train a T5-eQAG model directly on top of the summary using the annotated QAs, since T5-summary model already has knowledge on rich informative text.

To generate the questions and QA, as a text-to-text generation task, it is essential to model the joint conditional distribution $p(a, q|c)$, where (q, a, c) represents ground truth question, context passage and the answer, which is a span in c. We fine-tune a text-to-text transformer (T5) LM using a training samples D. We use maximum likelihood estimation (MLE) to train the T5 model using the θ parameters, where the objective is to minimizing the negative log-likelihood over training samples D:

$$\sum_{n=-1}^{|D|} \log P\theta(a^n, q^n|c^n) \qquad (2)$$

Given the input passage, the model jointly creates the question and the answer. The question tokens are created during sampling, then the response tokens. As a result, the creation of the answer is dependent on the input context (by focusing on the encoder) and the query (through self-attention in the decoder).

Most of the existing QA generation model facing an exposure bias issue. Therefore, we created a negative sample (questions, QAs) and trained an end-to-end T5_QAG model by introducing contrastive loss function. We trained two variants of t5 QG model. At first we fine-tuning the QA generation model on QA data (FairytaleQA and Hotpot QA) by minimizing the cross-entropy loss. In the next step QA model is trained on augmented data (both ground truth QAs and generated negative samples) with the contrastive loss and cross-entropy loss.

Contrastive Loss for T5 Model Fine Tuning. After training the t5 model for QAG task using cross entropy loss, in second step, we fine-tune the model for QAG task using the following loss function.

$$T_{loss} = Q_{task} + Q_{c\_loss} \tag{3}$$

where T_{loss} denotes the total loss, Q_{task} and Q_{c_loss} are the QA generation task loss and the contrastive losses.

$$Q_{c\_loss} = q_l * d_w^2 + (1 - q_l) * max(m - d_w, 0)^2 \tag{4}$$

where q_l is the ground-truth labels from our dataset, d_w is the Euclidean distance and m is the margin used for the contrastive loss function.

When the T5 model is trained with small amount of data, fine-tuning process with QA task loss is generally insufficient to achieve satisfactory performance. To address this issue, we generated the negative samples for each document d_i and fine-tune using both the data samples.

During fine-tuning, we pick one of the negative instances from the augmented data corpora (Ad) for each sample in the original data (Od) of size n at random. We generate the embedding $E_o, E_p \in R^{n*t*d}$, using the same T5 network as our encoder (En) up to a certain layer. Then, for each instance in each batch, we use a global average pooling (Gp) procedure to combine the vector of t tokens into a single vector of dimension d.

$$M_{c\_loss} = \frac{Ls_{qa\_rowQ} + Ls_{qa\_colQ}}{2} \tag{5}$$

As a result, the final embeddings O,P $\in R^{n*d}$ (O= Gp(En(Od)) and P = (Gp(En(Ad))) for the original and translated batches (Od and Ad) will be generated. We use the CLIP [36] approach to compute the contrastive loss with these n feature vectors. The logits matrix $Q \in R^{n*n}$ is obtained by multiplying the matrices O and P^T. The logits matrix Q is then subjected to the cross-entropy loss (Lce) row-by-row and column-by-column, with the diagonal locations serving as original classes for each row and column, respectively.

Relevancy Check Using BERT-Discriminator. For a given input text our QA model can generate a set of QAs. So, to determine if the generated QAs is pertinent to the input content, we fine tuned a BERT discriminator [8]). The discriminator is an binary classifier, uses the generated QAs q as well as the input document d to determine its relevance likelihood $p(q|d)$. Both FairytaleQA and the generated negative instances (QAs) are used to train the relevance discriminator. The ground-truth QAs q_i corresponds to the text is considered as the positive samples.

The negative sample q_j for a document-QA pairs (d, q_i) is generated using the following three ways: i) Negative instance q_j using random swapping: for the document d_i: we choose at random one of the ground truth QAs from another document d_j. ii) Negative instance q_j using inter-doc entity swapping: The entity in the ground-truth QAs q_i is swapped out for another entity of the same type that doesn't appear in the document d_i. The discriminator can use this to determine whether the query involves entities that aren't referenced in the text. iii) Negative instance q_j using intra-doc entity swapping: We also use a different entity from the same document in place of the one in the ground-truth QAs. This makes logical errors in the query.

For each ground-truth QAs, three negative samples are produced. We use the alpha-balanced focal loss [28] to resolve the imbalance between positive and negative samples, the focal loss is computed as:

$$F_l(P_i) = -\alpha(1 - (P_i)^\wp \log(P_i) \tag{6}$$

where P_i is the predicted probability for class i, $(1 - (P_i)^\wp)$ is a modulating factor with a tunable parameter $\wp \geq 0$ that smoothly adjusts the rate at which easy questions are down-weighted.

4   Experimental Results and Discussion

The proposed model generates potential question-answers to enhance the text content. We experimented with our strategy using textbooks from various subjects and grade levels.

In this work we trained T5 (Text-to-Text Transfer Transformer) base model from huggingface transformers[2]. The model receives the input sequence via input ids. The target sequence is provided to the decoder using the decoder input ids after being prefixed by a start-sequence token. EOS token is subsequently attached to the target sequence in teacher-forcing fashion, which correlates to the labels. The start-sequence token here is the PAD token.

4.1   Dataset

The FairytaleQA [48] data has 10,580 QA-pairs, which were drawn from 278 different novels, and each question has a label stating which narrative element(s) or relationship(s) it is intended to evaluate. We divided the dataset into train, validation, and test splits using 8,548/1,025/1,007 QA pairs.

We also used HotpotQA [49] which has 100K QA pairs on Wikipedia articles. It is challenging task to create a fluent, and pertinent, questions in HotpotQA, since it requires reasoning over multiple pieces of data in the text. We divide the data into 90,440 and 6,072 QAs as training, test samples, respectively. As the development set, we also hold aside 6,072 samples from the training data.

eBOOK data for testing eQA generation model: To test our QA generation model we used a collection of eBooks such as NCERT textbook (Science Grade X) and graduate levels textbooks - "Pattern Recognition and Machine Learning(PRML) (Bishop) [2].

Hyper-parameters for Fine Tuning T5-QG: For fine tuning the T5 model we used the AdaFactor optimizer and a maximum sequence length is set to 512. Weighted sampling based on language priors with exponent s = 0.7 is used to construct mini-batches. We follow the grid-search approach for choosing the best set of training parameters

---

[2] https://huggingface.co/docs/transformers/model_doc/t5.

(learning-rate(lr): $\{\{2,3,5\}e-3\}\}$ and batch size: $\{8,16,32,64\}$, warm-up ratio: 0.1), with AdaFactor optimizer. During the experiment, we found that a mini-batch size of 32 (lr: $3e-3$) shows decent results. We experimented with various pooling techniques (MEAN, and MAX). For the best results, we set MEAN as the default configuration.

The T5 eQA model is fine tuned for 1 to 4 epochs with max length = 512 and doc stride = 128. The best results were found at fourth epoch. Various max_length setup(such as 400), and other strides (192), were tried, but they didn't fare well with our cross-validation technique.

The generated questions (top k) are presented in Table 1, 2, 3, 4 and 5 for the test data from FairytaleQA dataset and PRML, NCERT CCT eBook. Table 1 present the generated questions for the sample input text drawn from PRML eBOOK [2] (ref. Fig. 2 for sample text). Table 2, and Table 3 highlight the generated questions for text from NCERT[3] CCT eBook, whereas Table 4 and Table 5 highlight the generated questions for sample text from FairytaleQA dataset [48]. Table 6 depict the sample input text from FairytaleQA data and the generated QAs by human experts (ground truth QAs), CBQA baseline, and our edu-QA generation model. The created QAs makes the students learning more effective, and assist them in improving their conceptual understanding ability. More examples (QAs) are presented in Appendix B.

1.2. Probability Theory

A key concept in the field of pattern recognition is that of uncertainty. It arises both through noise on measurements, as well as through the finite size of data sets. Probability theory provides a consistent framework for the quantification and manipulation of uncertainty and forms one of the central foundations for pattern recognition. When combined with decision theory, discussed in Section 1.5, it allows us to make optimal predictions given all the information available to us, even though that information may be incomplete or ambiguous.

Fig. 2. The input text were taken from PRML e-book (Sect. 1.2) by C. Bishop [2]. Students answer questions after reading text passages.

4.2 Model Evaluation

For automatic evaluation of the question-generation(QG) task, we use the BLEU scores[4] [33], Meteor [1] and Rouge-L measure[5] [27]. We validated the quality of the generated QAs on three experimental configuration. QA model trained i)only on HotpotQA, (ii)only on FairytaleQA data, iii)on both FairytaleQA and HotpotQA. The third

---

[3] https://ncert.nic.in/textbook.php.

[4] https://github.com/Tiiiger/bert_score.

[5] https://github.com/google-research/google-research/tree/master/rouge.

Table 1. The abstract summary and generated questions using our eQAG model(top k questions) for the input text from PRML eBook [2].

Text passage: A key concept in the field of pattern recognition is that of uncertainty....	
Abstract summary: Probability theory provides a consistent framework for the quantification and manipulation of uncertainty and forms one of the central foundations for pattern recognition....	
Generated top k questions using eQAG model (k=5)	
Q1	What does probability theory do?
Q2	What is one of the central foundations for pattern recognition?.
Q3	When are we equally likely to select any of the pieces of fruit in the box?
Q4	What is a consistent framework for the quantification and manipulation of uncertainty?
Q4	How much of the time do we pick the red box?

Table 2. The abstract summary and generated questions using our eQG model(top k questions) for the input text (Chap. 13) from NCERT CCT2 eBook.

Text passage: Use of innovative technologies like Silicon-On-Insulator (SOI), Complementary Metal-Oxide-Semiconductor (CMOS).capacitor-less memory, Micro-Optic-Electro-Mechanical-System (MOEMS), III-V compound materials-on-insulator and others have improved the performance and also reduced the size of consumer electronic devices. Apart from the development of innovative designs, researchers are also exploring materials other than Silicon...	
Abstract summary: Consumer electronics now perform better because to cutting-edge technologies like Silicon-On-Insulator (SOI),Complementary Metal-Oxide-Semiconductor -(CMOS) capacitor-less memory, Micro-Optic-Electro-Mechanical-System (MOEMS).....	
Generated top k questions using eQAG model (k=4)	
Q1	What does SOI stand for?
Q2	How much more efficient is graphene than silicon-based technology?
Q3	What is a form of pure carbon where a single layer of carbon atoms are arranged in a honeycomb lattice
Q4	What does MOEMS stand for?

Table 3. The abstract summary and generated questions using our eQG model(top k questions) for the input text (Chap. 13) from NCERT CCT2 eBook.

Text passage: A microprocessor is a multipurpose programmable logic device that reads binary instructions from a storage device called memory,...	
Abstract summary: Microprocessor,packages of multiple microchips and processors are relea--sed which fulfil and meet the performance....	
Generated top k questions using eQG model (k=3)	
Q1	What is a multiprocessing CPU?
Q2	What processor refers to two separate dualcore dies CPUs next to each other in one quadcore package?
Q3	What is the difference between a microprocessor and a Core CPU?

Table 4. The abstract summary and generated questions using our eQAG model(top k questions) for the FairytaleQA samples.

Text passage:
once upon a time there was a king who went forth into the world and fetched back a beautiful queen. and after they had been married a while god gave them a little daughter. Then there was great rejoicing in the city and throughout the country , for the people wished their king all that was good, since he was kind and just ....

Abstract summary: There there was a king who ventured into the world and brought back a lovely queen. The people then celebrated greatly in the city and across the nation, wishing their ruler nothing but the best....

Generated top k questions using eQAG model (k=5)	
Q1:	Who did the king bring back?
Q2	What did god give the king and queen after they had been married a while?
Q3	Who was appointed to keep an eye on the little princess?
Q4	How old was the child when she was told she must not be allowed out under the open sky?
Q5	Who entered the room while the child lay in its cradle?

Table 5. The abstract summary and generated questions using our eQAG model(top k questions) for the FairytaleQA samples.

Text passage:(FairytaleQA Dataset)
besides these, i want a cloak made of a thousand different kinds of skin, every animal in your kingdom must give a bit of his skin to it. but she thought to herself, this will be quite impossible, and i shall not ...
large forest. and as she was very much tired she sat down inside a hollow tree and fell asleep .

GTQ	Groud truth Q:what did the princess do while everyone was sleeping

Abstract summary: The king told the princess, "Every animal in your kingdom must contribute a ...
small coat of many skins, and covered her hands and face with soot.

Generated top k questions using eQAG model (k=5)	
Q1	What did the king tell the princess?
Q2	How many different kinds of skin did the princess want?
Q3	What items did the princess steal from her treasures?
Q4	'When was the day of the wedding announced?
Q5	What did the princess do while everyone slept?

Table 6. The generated QAs for the sample input text from FairytaleQA data. Here, the results of CBQA baseline model [25], our edu-QA generating mechanism, and the ground truth QAs are shown.

Text passage:(FairytaleQA Dataset)
besides these, i want a cloak made of a thousand different kinds of skin, every animal in your kingdom must give a bit of his skin to it . but she thought to herself, this will be quite impossible, ...
to god, and went out and travelled the whole night till she came to a large forest. and as she was very much tired she sat down inside a hollow tree and fell asleep .
Generated QAs using our edu-eQA and baseline model
Ground-Truth
Qst: what did the cloak have to made of ?   Ans: every animal   Qst: Where did the princess fall asleep ?   Ans: inside a hollow tree
CBQA Baseline (Patrick et al., 2021)[25]   Qst: How many different kinds of skin is a cloak made of?   Ans: a thousand
Top K QAs from our eQA model
Qst: What was the cloak made of?   Ans: a thousand pieces of fur   Qst: Where did the princess go after she commended herself to god?   Ans: a large forest   Qst: Why did the princess leave?   Ans: no chance of stopping her father's plans   Qst: What did the king say to the princess when the cloak was brought to him?   Ans: tomorrow shall be your wedding- day   Qst: What did the princess take from her treasures?   Ans: gold reel   Qst: What kind of tree did the princess fall asleep in?   Ans: hollow

setting, which shows a significant improvement over the preceding setups, was chosen as our base QA generation model(T5_eQAG) for further comparison to the earlier work. Results are shown in Table 7. We see that the model optimised on FairytaleQA alone shows significant improvement over prior works. This is due to the disparities in domain and distribution between the two data sets.

Table 7. Comparison of our contrastive edu-QAG models with various experimental parameters on the FairytaleQA dataset.

QG model	Eval metric: Rouge-L	
	Validation_data	Test_data
T5$_{base}$ fine-tuned on HotpotQA	0.408	0.431
T5$_{base}$ fine-tuned on FairytaleQA	0.519	0.536
T5$_{base}$ fine-tuned on both HotpotQA and FairytaleQA	0.514	0.522

4.3 Comparison with Baselines

Eval of QG model: We evaluate our contrastive trained edu-question generation(eQG) model against baseline QG model. Table 8 provides an overview of the technologies used by each model as well as the performance outcomes. Our model achieves a comparable BLEU4 with the baseline QG model in HotpotQA without using the answer information or any external linguistic knowledge. This illustrates the effectiveness of contrastive training of language model for the educational question generation task.

Eval of QAG model: We examined the generated QAs relevance to the input text and their similarity to ground-truth QA-pairs given the same text passage. For each pair of QAs we concatenate the question, answer while computing the Rouge-L precision score. We used ranking metric to examine an upper bound (N) of generated QAs for the given input text passage. Here we examine how relevant the generated QAs are to the input text and how similar the generated QAs are to the ground-truth QAs. To assess the QAs generation task, we employed MAP@K ($k = 1, 2, 3, \ldots, N$). There are typically three QAs per text passage in FairytaleQA data, so we used MAP@3 as a relevant metric for smoother comparison. Table 9 presents the comparison results for QA generation task on FairytaleQA data. On test splits of the FairytaleQA dataset, our eQA model outperforms the baseline model [25] with noticeably improved Rouge-L precision performance. However, the Rouge-L is unable to assess the syntactic and semantic quality of generated QAs. We also undertake a human evaluation, to provide an realistic evaluation on semantic quality of the generated QAs.

We compared our T5-summarizing module against several summarising techniques in order to examine its effects. The summarising techniques comprise: 1) Lead_3(Top3)- our QA generation model is fed with the first three sentences of a paragraph as the summary; 2) last_3(Bottom3)- last three sentences in a paragraph to serve as the summary. 3) Random3- sentences were chosen at random from a paragraph and feed them into the QA generation model. The findings are presented in Table 10, which demonstrates that selecting the phrases from a paragraph does not adequately capture important events for the creation of educational QAs. The rich informative sentences or events can be effectively extracted using our summary generation method. Although using all of the sentences can increase accuracy at the cost of memory, the overall F1 score is still not very high.

Table 8. The comparison results of our model with prior work for question generation task on HoTPoT QA dataset. (BS1: BLEU score 1, BS4: BLEU score 4

QG model	Eval metrics			
	BS1	BS4	Meteor	Rouge-L
Deep QG [31]	40.55	15.53	20.15	36.94
Reinforcement learning QG [31]	37.97	15.41	19.61	35.12
Our model(Contrastive_T5_QG)	41.28	18.22	20.37	43.61

Table 9. The comparison of SOTA model with our model for QA generation task on FAIRQA dataset(Eval_metric: MAP@K).

QA model	Eval metric:MAP@k		
	k = 1	k = 3	k = 5
CBQA [25]	0.257	0.384	0.436
Contrastive FT_T5_QAG	0.322	0.479	0.531

Table 10. The comparison of several summarising techniques results on FAIRYQA dataset (Eval parameter: Rouge-L on eQAG).

Summary method	Precision	Recall	F1 score
First_3 sent	23.52	28.30	22.47
Last_3 sent	22.31	27.52	22.28
Random_3 sent	21.37	26.53	21.19
Our T5_sum	33.06	29.91	27.54

4.4 Human Evaluation for Relevancy Testing

We conducted a human evaluation to validate the appropriateness of the generated QAs on eBook content. We invited five students to assess the quality of the QAs that were generated. We selected 20 text documents from the test set of PRML and NCERT books at random to facilitate the evaluation procedure. Our model generates a set of QAs for each text document (total 120 QAs for 20 text document). Evaluators were asked to review the generated QAs using the evaluation criteria of "relevancy" and "readability". The "relevance" factors is the one that is most closely tied to education; the readability parameter assess factual accuracy and fluency. "Readability": check for the syntax and semantics (English grammar and words) of the generated QAs. "Relevance of the Question": how relevant the generated question is to the input text. "Relevance of the Answer": how relevant the generated answer to the question. The evaluators rate the QAs on a five-point Likert-scale. The generated QAs receives a decent Krippendoff's alpha [16] ratings readability: 0.76, QA_relevancy: 0.71 indicating an acceptable consistency [15].

To assess how well our model performed in comparison to the baseline model, we used t-tests. Table 11 displays the outcomes of the human evaluation for the QA task. Our model(mean vale:4.73, stdv:0.68) outperformed the CBQA model [25] in terms of the QA-Readability parameter (mean value:4.01, stdv:1.18, t (477):7.33, p < 0.01) by a large margin. For the Relevancy of the generated question, our model shows mean value: 4.41, stdv: 1.10 and shows marginal improvement over the baseline (mean value: 4.09, stdv: 1.27, t (477): 1.98, p < 0.05). The outcome demonstrates that our model can produce more pertinent questions to the input text than the baseline model.

Regarding the "Relevancy" of the generated answer, we take into account whether or not the generated answer suits the generated question. The performance of our model (mean value: 3.95, stdv: 1.24) was better than that of the CBQA model [25](mean value: 3.57, stdv: 1.51, t (477): 0.58, p:0.56) demonstrating the effectiveness of the contrastive loss on the training process. The results prove that the generated QAs are better fit the educational scenario.

Table 11. The results of human evaluation for QA relevancy testing sample e-Book content (Eval parameter: mean and stdv).

Eval_parameter (scale: 1 to 5)	CBQA		Our T5_QA	
	mean	stdv	mean	stdv
QA_Readability	4.01	1.18	4.73	0.68
Relevance of the Question	4.09	1.27	4.41	1.10
Relevance of the Answer	3.57	1.51	3.95	1.24

5 Conclusion and Future Scope

In this work, we explore the education Question-Answer generation task (eQA) for e-book content to create an effective learning platform for curious learner. The work has two phases- "Text summarizer", followed by "Educational Question-Answer generation" using T5 transformer model. We introduced a contrastive loss for training the Text-to-Text Transformer (T5) eQA model. Experiments on FairytaleQA, HotPot QA, and NCERT e-book shows that our model succeeds to produces relevant QAs at scale. To examine relevancy of the generated QAs, we further built an BERT discriminator, which analyze how relevant the generated QAs inline with the input document. Our research establishes a strong foundation for the hopeful future of automating educational QG and QA processes using AI. The possible future direction could be design a context-aware QG and QA model, where the generation of a new text (QA or QG) is conditioned on previous generations as well as the key topics of the eBook.

Appendix

A BERT Score for Semantic Match

Besides the standard evaluation metrics such as, METEOR and Rouge-L scores, we also employ BERTScore [50] to assess the semantic similarity of questions generated by our QG model with the ground-truth questions, and present the average precision, recall, and F1 values. Table 12 outlines the computed BERT scores for top k generated questions.The results show that our model can generate relevant questions for the given text.

Table 12. The comparison results (BERT score) of our model when k = 1, 3, 5 (top k questions) on FairytaleQA test data.

QG model	BERT score on test set		
	Precision	Recall	F1
$T5_{base}$: eQG (k = 1)	0.842	0.813	0.827
$T5_{base}$: eQG (k = 3)	0.865	0.849	0.856
$T5_{base}$: eQG (k = 5)	0.884	0.872	0.878

B Few More Examples of Generated QAs for Text Document

Two more samples (input text), the ground truth QAs, generated QAs by CBQA model, and our edu-QA generation model. Tables 13 depict the sample text from NCERT CCT eBook, generated summary, QAs by our model and generated QAs by CBQA model [25]. Table 14 contain a generated QAs for the sample input text from FairytaleQA dataset, associated ground truth QAs, and generated summary, QAs by our model and generated QAs by CBQA model [25].

Table 13. The generated QAs for the sample input text from NCERT e-BOOK (Subject: CCT-Chap. 13) data. Here, the results of PAQ baseline model, our edu-QA generating mechanism, and the ground truth QAs are shown.

Text passage:NCERT e-BOOK (Subject: CCT- Chapter 13)
Use of innovative technologies like Silicon-On-Insulator (SOI) Complementary Metal-Oxide-Semiconductor (CMOS) capacitor less memory, Micro-Optic-Electro-Mechanical-System (MOEMS), compound materials-on-insulator and others have improved the performance and also reduced the size of consumer electronic devices. Apart from the development of innovative designs, researchers are also exploring materials other than Silicon to develop high speed electronic chips. n One such example is Graphene, which is a form of pure carbon where a single layer of carbon atoms are arranged in a honeycomb lattice. It could allow electrons to process information and produce radio transmissions 10 times better than silicon-based devices. Usage of Graphene will produce faster and more powerful cell phones, computers and other electronic devices, since the mobility of electrons is more than any existing semiconductor materials like Silicon, GaAs and carbon nano tubes......
Abstract summary:
Consumer electronics now perform better because to cutting-edge technologies like Silicon-On-Insulator (SOI), Complementary Metal-Oxide-Semiconductor (CMOS) capacitor-less memory, Micro-Optic-Electro-Mechanical-System (MOEMS), compound materials-on-insulator, and others. To create high-speed electronic devices, researchers are also looking into alternatives to silicon. One such instance is the pure carbon material known as graphene, which consists of a single layer of carbon atoms organised in a honeycomb lattice.
CBQA Baseline (Patrick et al., 2021)[25]
Qst: What is a form of pure carbon where a single layer of carbon atoms are arranged in a honeycomb lattice? Ans: Graphene
Generated Top K QAs from our eQA model
Qst: What does SOI stand for? Ans: Silicon-On-Insulator Qst: What is the name of the innovative technologies that have improved n the performance and reduced the size of consumer electronic devices? Ans: SiliconOn-Insulator SOI and Complementary Metal-Oxide-Semiconductor CMOS Qst: What are researchers exploring materials other than Silicon to develop? Ans: High speed electronic chip Qst: What does MOEMS stand for? Ans: Micro-Optic-Electro-Mechanical-System Qst: How much more efficient is graphene than silicon-based technology? Ans: ten times more effectively

Table 14. The generated QAs for the sample input text from FairytaleQA [48] data. Here, the results of CBQA baseline model (Patrick et al., 2021) [25], our edu-QA generating mechanism, and the ground truth QAs are shown.

Text passage:(FairytaleQA Dataset) Once upon a time there was a king who went forth into the world and fetched back a beautiful queen. and after they had been married a while god gave them a little daughter. then there was great rejoicing in the city and throughout the country, for the people wished their king all that was good, since he was kind and just . while the child lay in its cradle, a strange - looking old woman entered the room, and no one knew who she was nor whence she came. the old woman spoke a verse over the child, and said that she must not be allowed out under the open sky until she were full fifteen years of age, since otherwise the mountain troll would fetch her. when the king heard this he took her words to heart, and posted guards to watch over the little princess so that she would not get out under the open sky ...
Abstract summary: There there was a king who ventured into the world and brought back a lovely queen. The people then celebrated greatly in the city and across the nation, wishing their ruler nothing but the best. An elderly woman read a poetry to the kid as it was still in its cradle and warned that she shouldn't be allowed outside until she was at least fifteen years old for fear that the mountain troll would come and fetch her. When the monarch heard this, he took her advice to heart and appointed guards to keep an eye on the little princess to prevent her from escaping into the open sky.
Ground-Truth Qst: what type of ruler was the king ? Ans: kind and just.
CBQA Baseline (Patrick et al., 2021)[25] Qst: What did the king fetch back? Ans: beautiful queen
Generated Top K QAs from our eQA model Qst: Who did the king bring back? Ans: a beautiful queen   Qst: What did god give the king and queen after they had been married a while? Ans: a little daughter   Qst: Who entered the room while the child lay in its cradle? Ans: a strange - looking old woman   Qst: How old was the child when she was told she must not be allowed out under the open sky? Ans: fifteen years   Qst: What did the king post to watch over the little princess? Ans: posted guards to watch over the little princess so that she would not get out under the open sky.   Qst: What was there in the city and throughout the country after the king gave his daughter? Ans: great rejoicing

References

1. Banerjee, S., Lavie, A.: METEOR: an automatic metric for MT evaluation with improved correlation with human judgments. In: Proceedings of the ACL Workshop on Intrinsic and Extrinsic Evaluation Measures for Machine Translation and/or Summarization, Ann Arbor, Michigan, June 2005, pp. 65–72. Association for Computational Linguistics (2005)
2. Bishop, C.M.: Pattern Recognition and Machine Learning (Information Science and Statistics). Springer, Heidelberg (2006)
3. Chan, Y.-H., Fan, Y.-C.: A recurrent BERT-based model for question generation. In: Proceedings of the 2nd Workshop on Machine Reading for Question Answering, Hong Kong, China, November 2019, pp. 154–162. Association for Computational Linguistics (2019)
4. Chen, G., Yang, J., Gasevic, D.: A comparative study on question-worthy sentence selection strategies for educational question generation. In: Isotani, S., Millán, E., Ogan, A., Hastings, P., McLaren, B., Luckin, R. (eds.) AIED 2019. LNCS (LNAI), vol. 11625, pp. 59–70. Springer, Cham (2019). https://doi.org/10.1007/978-3-030-23204-7_6
5. Cho, W.S., et al.: Contrastive multi-document QG. In: Proceedings of the 16th European Conference, Chapter of the ACL, pp. 12–30 (2021)
6. De Kuthy, K., Kannan, M., Ponnusamy, H.S., Meurers, D.: Towards automatically generating questions under discussion to link information and discourse structure. In: Proceedings of the 28th International Conference on Computational Linguistics, pp. 5786–5798 (2020)
7. Demszky, D., Guu, K., Liang, P.: Transforming question answering datasets into natural language inference datasets. CoRR, abs/1809.02922 (2018)
8. Devlin, J., Chang, M.-W., Lee, K., Toutanova, K.: BERT: pre-training of deep bidirectional transformers for language understanding. In: Proceedings of the 2019 Conference of the North American Chapter of the Association for Computational Linguistics: Human Language Technologies, (Long and Short Papers), Minneapolis, Minnesota, June 2019, vol. 1, pp. 4171–4186. Association for Computational Linguistics (2019)
9. Dhole, K.D., Manning, C.D.: Syn-QG: syntactic and shallow semantic rules for question generation. arXiv, abs/2004.08694 (2020)
10. Dong, L., et al.: Unified Language Model Pre-training for Natural Language Understanding and Generation. Curran Associates Inc., Red Hook (2019)
11. Du, X., Shao, J., Cardie, C.: Learning to ask: neural question generation for reading comprehension. CoRR, abs/1705.00106 (2017)
12. Ebersbach, M., Feierabend, M., Nazari, K.B.B.: Comparing the effects of generating questions, testing, and restudying on students' long-term recall in university learning. Appl. Cogn. Psychol. **34**(3), 724–736 (2020)
13. Kurdi, G., Leo, J., Parsia, B., Sattler, U., Al-Emari, S.: A systematic review of automatic question generation for educational purposes. Int. J. Artif. Intell. Educ. **30**, 121–204 (2020)
14. Golinkoff, R.M.: Language matters: denying the existence of the 30-million-word gap has serious consequences. Child Dev. **90**(3), 985–992 (2019)
15. Gretz, S., Bilu, Y., Cohen-Karlik, E., Slonim, N.: The workweek is the best time to start a family - a study of GPT-2 based claim generation. In: Findings of the Association for Computational Linguistics, EMNLP 2020, November 2020. Association for Computational Linguistics, pp. 528–544 (2020)
16. Gwet, K.L.: On the Krippendorff's alpha coefficient (2011)
17. Ganotice, F.A., Jr., Downing, K., Mak, T., Chan, B., Lee, W.Y.: Enhancing parent-child relationship through dialogic reading. Educ. Stud. **43**(1), 51–66 (2017)
18. Kočiský, T., et al.: The NarrativeQA reading comprehension challenge. Trans. Assoc. Comput. Linguist. **6**, 317–328 (2018)

19. Krishna, K., Iyyer, M.: Generating question-answer hierarchies. In: Proceedings of the 57th Annual Meeting of the ACL, Florence, Italy, pp. 2321–2334, July 2019. Association for Computational Linguistics (2019)
20. Kuamr, S., Chauhan, A.: Augmenting textbooks with CQA question-answers and annotated YouTube videos to increase its relevance. Neural Process Lett. (2022)
21. Kumar, S., Chauhan, A.: Enriching textbooks by question-answers using CQA. In: 2019 IEEE Region 10 Conference (TENCON), TENCON 2019, pp. 707–714 (2019)
22. Kumar, S., Chauhan, A.: Recommending question-answers for enriching textbooks. In: Bellatreche, L., Goyal, V., Fujita, H., Mondal, A., Reddy, P.K. (eds.) BDA 2020. LNCS, vol. 12581, pp. 308–328. Springer, Cham (2020). https://doi.org/10.1007/978-3-030-66665-1_20
23. Kumar, S., Chauhan, A.: A finetuned language model for recommending cQA-QAs for enriching textbooks. In: Karlapalem, K., et al. (eds.) PAKDD 2021. LNCS (LNAI), vol. 12713, pp. 423–435. Springer, Cham (2021). https://doi.org/10.1007/978-3-030-75765-6_34
24. Labutov, I., Basu, S., Vanderwende, L.: Deep questions without deep understanding. In: Proceedings of the 53rd Annual Meeting of the Association for Computational Linguistics and the 7th International Joint Conference on Natural Language Processing (Volume 1: Long Papers), Beijing, China, July 2015, pp. 889–898. Association for Computational Linguistics (2015)
25. Lewis, P., et al.: PAQ: 65 million probably-asked questions and what you can do with them. Trans. Assoc. Comput. Linguist. **9**, 1098–1115 (2021)
26. Li, M., et al.: Timeline summarization based on event graph compression via time-aware optimal transport. In: Proceedings of the 2021 Conference on Empirical Methods in Natural Language Processing, Punta Cana, Dominican Republic, pp. 6443–6456, November 2021. Association for Computational Linguistics (2021)
27. Lin, C.-Y.: ROUGE: a package for automatic evaluation of summaries. In: Text Summarization Branches Out, Barcelona, Spain, July 2004, pp. 74–81. Association for Computational Linguistics (2004)
28. Lin, T.-Y., Goyal, P., Girshick, R., He, K., DollÁr, P.: Focal loss for dense object detection. IEEE Trans. Pattern Anal. Mach. Intell. **42**(2), 318–327 (2020)
29. Lyu, C., Shang, L., Graham, Y., Foster, J., Jiang, X., Liu, Q.: Improving unsupervised question answering via summarization-informed question generation. In: Proceedings of the EMNLP, pp. 4134–4148 (2021)
30. Nguyen, T., et al.: MS MARCO: a human generated machine reading comprehension dataset. CoRR, abs/1611.09268 (2016)
31. Pan, L., Xie, Y., Feng, Y., Chua, T.-S., Kan, M.-Y.: Semantic graphs for generating deep questions. In: Proceedings of the 58th Annual Meeting of the Association for Computational Linguistics, July 2020, pp. 1463–1475. Association for Computational Linguistics (2020)
32. Pan, L., Xie, Y., Feng, Y., Chua, T.-S., Kan, M.-Y.: Semantic graphs for generating deep questions. In: Proceedings of the 58th Annual Meeting of the ACL, pp. 1463–1475 (2020)
33. Papineni, K., Roukos, S., Ward, T., Zhu, W.-J.: Bleu: a method for automatic evaluation of machine translation. In: Proceedings of the 40th Annual Meeting on Association for Computational Linguistics, ACL 2002, USA, pp. 311–318. Association for Computational Linguistics (2002)
34. Pyatkin, V., Roit, P., Michael, J., Tsarfaty, R., Goldberg, Y., Dagan, I.: Asking it all: generating contextualized questions for any semantic role. CoRR, abs/2109.04832 (2021)
35. Qi, W., et al.: ProphetNet: predicting future n-gram for sequence-to-sequence pre-training. In: EMNLP 2020, pp. 2401–2410 (2020)
36. Radford, A., et al.: Learning transferable visual models from natural language supervision. CoRR, abs/2103.00020 (2021)
37. Radford, A., Wu, J., Child, R., Luan, D., Amodei, D., Sutskever, I.: Language Models are Unsupervised Multitask Learners (2019)

38. Ruan, S., et al.: Quizbot: a dialogue-based adaptive learning system for factual knowledge. In: Proceedings of the 2019 CHI Conference on Human Factors in Computing Systems, CHI 2019, pp. 1–13. Association for Computing Machinery, New York (2019)

39. Rush, A.M., Chopra, S., Weston, J.: A neural attention model for abstractive sentence summarization. In: Proceedings of the 2015 Conference on Empirical Methods in Natural Language Processing, Lisbon, Portugal, September 2015, pp. 379–389. Association for Computational Linguistics (2015)

40. Scialom, T., Piwowarski, B., Staiano, J.: Self-attention architectures for answer-agnostic neural question generation. In: Proceedings of the 57th Annual Meeting of the Association for Computational Linguistics, Florence, Italy, July 2019, pp. 6027–6032. Association for Computational Linguistics (2019)

41. Stasaski, K., Hearst, M.A.: Multiple choice question generation utilizing an ontology. In: Proceedings of the 12th Workshop on Innovative Use of NLP for Building Educational Applications, Copenhagen, Denmark, September 2017, pp. 303–312. Association for Computational Linguistics (2017)

42. Tuan, L.A., Shah, D.J., Barzilay, R.: Capturing greater context for question generation. CoRR, abs/1910.10274 (2019)

43. Wang, D., Liu, P., Zheng, Y., Qiu, X., Huang, X.: Heterogeneous graph neural networks for extractive document summarization. In: Proceedings of the 58th Annual Meeting of the Association for Computational Linguistics, July 2020, pp. 6209–6219. Association for Computational Linguistics (2020)

44. Wang, T., Yuan, X., Trischler, A.: A joint model for question answering and question generation. In: Learning to Generate Natural Language Workshop, ICML 2017 (2017)

45. Wang, Z., Lan, A.S., Nie, W., Waters, A.E., Grimaldi, P.J., Baraniuk, R.G.: QG-Net: a data-driven question generation model for educational content. In: Proceedings of the Fifth Annual ACM Conference on Learning at Scale, L@S 2018. Association for Computing Machinery, New York (2018)

46. Xu, J., Gan, Z., Cheng, Y., Liu, J.: Discourse-aware neural extractive text summarization. In: Proceedings of the 58th Annual Meeting of the Association for Computational Linguistics, pp. 5021–5031, July 2020. Association for Computational Linguistics (2020)

47. Xu, Y., Wang, D., Collins, P., Lee, H., Warschauer, M.: Same benefits, different communication patterns: comparing children's reading with a conversational agent vs. a human partner. Comput. Educ. (2021)

48. Xu, Y., et al.: Fantastic questions and where to find them: FairytaleQA - an authentic dataset for narrative comprehension. Association for Computational Linguistics (2022)

49. Yang, Z., et al.: HotpotQA: a dataset for diverse, explainable multi-hop question answering. CoRR, abs/1809.09600 (2018)

50. Zhang, T., Kishore, V., Wu, F., Weinberger, K.Q., Artzi, Y.: BERTscore: evaluating text generation with BERT. CoRR, abs/1904.09675 (2019)

51. Zhao, Y., Ni, X., Ding, Y., Ke, Q.: Paragraph-level neural question generation with maxout pointer and gated self-attention networks. In: Proceedings of the 2018 Conference on Empirical Methods in Natural Language Processing, Brussels, Belgium, October–November 2018, pp. 3901–3910. Association for Computational Linguistics (2018)

Data Science: Applications

A Machine and Deep Learning Framework to Retain Customers Based on Their Lifetime Value

Kannan Kumaran[1]([envelope])[iD], Pramod Pathak[2][iD], Rejwanul Haque[1][iD], and Paul Stynes[1][iD]

[1] National College of Ireland, Dublin 1 D01 K6W2, Ireland
yeshkan55555@gmail.com
[2] Technical University Dublin, Grangegorman Lower, Smithfield, Dublin 7, Ireland

Abstract. Customer Lifetime Value (CLV) measures the average revenue generated by a customer over the course of their association with the firm. The Recency Frequency Monetary (RFM) Model is used to calculate the CLV. Recency is the latest item purchased. The number of times an item is purchased is the Frequency. Monetary is the price spent on the product by customers. CLV is measured using previous customer transactions of RFM factors. This research proposes a Deep Learning Customer Retention Framework to predict the Customer Lifetime Value in order to retain customers through an effective Customer Relationship Management strategy. The proposed framework combines clustering and regression models to analyze the significant variables for predicting the lifetime value of customers. Customers are categorized into levels such as high medium and low profitable customers based on their lifetime value. This research compares Deep Neural Network models, Machine Learning models and Probabilistic models. The Deep Neural Network is ANN. The machine learning models are Linear Regression, Random Forest, Gradient Boosting. The probabilistic models are Gamma-Gamma and Betageometric/negative binomial. The models are compared in order to predict the level of profitable customers. Results demonstrate that Deep Neural Network (DNN) model outperforms the other models with 71% accuracy. Improved prediction model for CLV and segmentation assists the firms to plan and decide relevant CRM strategies such as customer profitability analysis, cross-selling and one to one marketing for the future.

Keywords: Customer lifetime value · Recency Frequency Monetary (RFM) · Customer retention · Deep neural network

1 Introduction

Customer Lifetime Value (CLV) is the predicted future revenue to the firm over the lifetime of a customer. Customer Relationship Management (CRM) is the process of offering the relevant product or service to the right customer at the right time through the right medium [1]. CLV is a concept in CRM to establish

© The Author(s), under exclusive license to Springer Nature Switzerland AG 2022
P. P. Roy et al. (Eds.): BDA 2022 India, LNCS 13773, pp. 91–103, 2022.
https://doi.org/10.1007/978-3-031-24094-2_6

a long-term relationship between the customer and a firm. The calculation of CLV enables the firm to determine the monetary value of each consumer thereby guiding how much a marketing department should be ready to spend to obtain that customer. It is a direct marketing idea that considers long-term customer behaviour to be the key to success. This could be accomplished by creating a machine & deep learning model which can recognize the buying patterns. CLV predictions are made by the model trained with previous transactional data [2]. Buying patterns help to identify high, medium and low profitable customers. Customer segmentation is used to classify customer groups such as high profitable based on similarities such as purchasing premium products. Due to numerous transactions, customer group classification is hard to attain using traditional practices. Because of this customer segmentation utilizing ML clustering is used in this research to categorize customers as high, medium, and low profitable categories [3]. This divides the customers into many clusters based on their shared qualities.

The aim of this research is to investigate to what extent a Deep Learning Customer Retention Framework can retain Customers based on their Lifetime Value. The major innovative contribution of this research is a Customer Retention framework that combines Deep Neural Network (DNN) and Customer Lifetime Value Prediction Models in order to identify profitable customers that need to be retained. A minor contribution of this research is a comparison of Deep and ML models with probabilistic methods. To identify the optimum prediction model this research compares Deep Neural Network, Linear Regression, Random Forest, Gradient Boosting gamma-gamma and Beta Geometric/Negative Binomial. The most accurate model was selected based on the evaluation metrics accuracy ($R2$ score), Mean Absolute Error (MAE) and Mean Squared Error (MSE).

2   Related Work

Customer Lifetime Value (CLV) helps in retaining the existing customers and also to acquire new customers. Customer Lifetime Value has been researched using machine learning and ML segmentation.

Machine learning has been used for CLV to provide recommendations in developing a marketing plan [4]. The aim of this project is to examine the company's customer sales data and forecast the customer lifetime value. Also, Customer segmentation is done to develop focus on groups based on the predicted CLV. Based on machine learning models that predict CLV and segmentation, this paper guides in future planning and decision-making marketing strategies. Gradient Boosting Regressor model outperforms other ML models with 84% accuracy. Through segmentation it is inferred that nearly half of the firm's value is contributed by 17% of customers. This research recommends the usage of Boosting algorithms for attaining better predicted outcomes.

Customer Lifetime Value Model Framework Using Gradient Boost Trees with RANSAC Response Regularization was performed to offer a mathematical multilayer model framework for calculating customer lifetime value based on a rigorous theoretical taxonomy as well as assumptions grounded on client characteristics [5]. Rather than using a gradient boosting technique to directly boost a base learner, this work uses RANSAC regularization to boost weak learners. The results are compared to a gradient boost with lasso regularization fitted by the complete training set and tested on a genuine customer base. In severely skewed customer data, experimental assessment reveals that the suggested framework delivers an accuracy of 80% with 0.12 Mean Square Error (MSE) when compared to another evaluated method. From a technical and business standpoint, there is still a lot of improvisations such as reducing the MAE and customer behavioral segmentation which will be attempted to resolve in current research.

CLV modelling assists to determine a customer's expected business value and enables firms to allocate resources efficiently in their businesses. The authors [2] predicted Customer Class using Customer Lifetime Value with Random Forest Algorithm for the following year based on their CLV, which will assist the online retailer in determining which clients should be engaged for long-term CRM. The Model is trained using the Random Forest (RF) method, and Random Search tuning is used to get the greatest prediction accuracy. On the same data set, an experimental study is done to compare with the AdaBoost method. Models using the ideal hyperparameter value from Random Search out perform AdaBoost models in terms of accuracy. The model's accuracy with default hyperparameters and all features is 81.46% with improvement to (82.26%) when only the selected characteristics from feature selection were used. The probability of a Random Forest model with the best hyperparameters adjusted by Random Search rose by 2%. Product interest will be advised based on the client class and preferences, which will lead to a rise in sales and assist to establish a stronger relationship with potential consumers, also motivate low class customers to enhance retention strategy.

Customer segmentation using machine learning was carried out analysing data from previous transactions to divide clients into several groups, ranging from high to low valuable. In customer relationship management, the RFM model is an essential quantitative analytical model. Research on improved RFM customer segmentation models, based on k-means algorithm, differentiated customers depending on various cluster groups in order to produce an advanced RFM model. The authors [6] acquired data to create parameters such as R, F, M, and C. C is a newly added parameter that identifies consumers who placed orders at the same time and belong to the same cluster group. The data was standardized after the RFMC values were calculated. The elbow technique was used to find the best cluster and was found to be 5. Both the traditional RFM and the enhanced RFMC model have been clustered. Finally, both approaches yielded similar results, with high recency values. The author performed well on the advanced RFMC model but a snake plot to highlight the variations between the models could have been made. For this reason, the proposed model would be implemented in current research to obtain better customer segmentation.

Customers are grouped according to their characteristics. The information from the various clusters will finally reveal which group the new customer will belong to, with the mean value serving as the primary indicator using three clustering approaches namely k-means, agglomerative, and mean shift [5]. As part of the data pre-processing, data scaling was performed. The elbow technique was applied in the k-means clustering method, and the best cluster was found to be 5. The silhouette score assessment technique was used to evaluate the outcomes of clustering techniques, and it was discovered that clustering methods (Agglomerative & K-means) perform better (0.56) than mean shift method (0.52). The author incorporated additional models and evaluation methods to obtain better clustering which provides reason to perform comparisons in the research. The different evaluation techniques used in this paper for comparison promises to identify the optimum clustering model.

In conclusion, the state of the art could be improved with additional data points and hyperparameters for better model training. Although Deep Neural Network using ANN has already been experimented to retain customers, the current research compares ML & DNN considering customer segmentation model. Also, the state of art with regards to customer segmentation indicates usage of RFM factors for classification of customer profitable class. This research proposes a Deep Learning integrated CLV along with ML clustering to guide effective CRM strategy. DNN will be identified as optimal model compared to BG/NBD, Gamma-Gamma, Linear Regression, Random Forest and Gradient Boosting. K-means clustering will be identified as better classifier compared to Hierarchical and manual RFM segmentation.

3   Methodology

The research methodology consists of five steps namely data understanding, data preparation, data transformation, data modelling, evaluation and results as shown in Fig. 1.

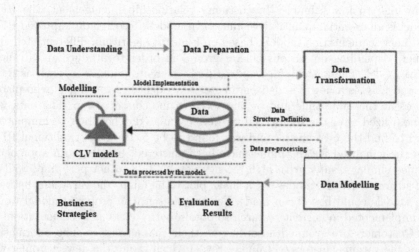

Fig. 1. Research methodology

The first step, Data Understanding involves fetching of data from UCI machine learning repository to perform the research. A UK based online retail data consisting of transactions between 01/12/2009 and 09/12/2011 was collected. There are 541910 on year (2010–2011) records of customer transactions having 8 features namely customer ID, Invoice, stock code, description, invoice, date, price, quantity and country [8].

The second step, Data Preparation involves selection of variables from the dataset such as Invoice, StockCode, Quantity, Price, InvoiceDate and Customer ID. Creation of an aggregated field named Total Price defining total price spent per product in each transaction obtained by multiplying Quantity with Price. Division of the variable InvoiceDate into two variables InvoiceDate and Invoice-Day that helps distinguish different transactions by same customer at different times on the same day. The transactions having missing values are filtered out with regards to Customer ID. To avoid incorrect CLV and customer classification predictions data cleaning steps such as outlier identification and removal, duplicate elimination and value scaling was done.

The third step, Data Transformation involves generation of recency, frequency, and monetary parameters for each customer ID. This is because the research aims to predict CLV based on RFM model. To generate the RFM, the customer id was grouped using the group-by function and aggregated with invoice date, invoice number, and total sales amount. The scores are provided as input for customer lifetime value prediction and clustering model. To construct the recency values in invoice dates, a lambda function was built using the differentiation between last purchase and recent date of the customer. The count of invoice numbers for each customer was acquired using the count() method to frame the frequency values. To construct the monetary values, the sum of the sales amounts for each customer was acquired using the sum() function. Thus, classification of high, medium and low customers was possible with RFM calculation. The data set was split into a ratio of (80:20) for training and validation.

The fourth step, Data Modelling involves model training, model conversion, and model implementation. The data has been modelled for CLTV prediction and customer segmentation. For this goal, both statistical and machine learning approaches are employed. Statistical methods such as beta geometric and gamma-gamma were used. Gradient Boosting and Deep Neural Network models were trained to learn and forecast continuous values. The predictor variable is the calculated CLV, and the model is trained on the training set and then validated on the test set in order to understand the model accuracy. Customer Segmentation is achieved by ML clustering algorithms such as K-means and Hierarchical clustering to group customers into different clusters based on their shared purchasing behaviour.

The fifth step, Evaluation and Results involves evaluating the performance of each of the deep and machine learning models using Accuracy (R2), Mean Squared Error, Mean Absolute Error and Snake Plot. Snake plot was used in this study to compare RFM groups generated using Machine Learning clustering to the actual RFM groups through visualizations for easy identification. The metrics were utilized to evaluate the models of CLV prediction. The optimal model with the best fit will have higher accuracy and low error score.

The sixth step, Business Strategies can be adopted based on the results of our study. It supports to offer Value-Oriented Content, Discounts at the Right Time, Upselling and Cross-Selling Strategies, Top-Notch Customer Service and get most out of data driven marketing.

4   Design Specification

A Deep Learning Customer Retention Framework architecture combines a deep neural network and ML clustering as shown in Fig. 2. Components of Customer Segmentation Models are discussed in Sect. 4.1. Components of Customer Lifetime Value Prediction Models are discussed in Sect. 4.2.

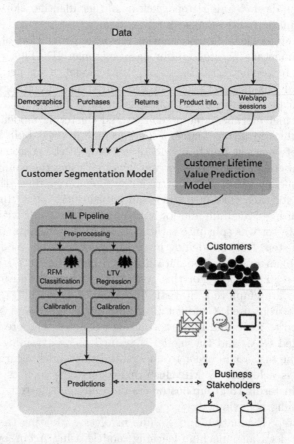

Fig. 2. Design of CLV prediction and segmentation

4.1   Customer Segmentation Models

The customer groups ranging from profitable to non-profitable can be formed by the execution of RFM parameters based on the previous transactions of customer. The recency parameter indicates how recently a customer made a purchase. The invoice and current date will be used to measure this. Frequency refers to the number of times a customer makes a purchase in a certain time period. The frequency will be determined using the customer ID. The entire amount spent by the customer in all transactions is referred to as monetary. Total Price and customer ID is used to measure this factor.

K-means clustering is used to segment customers with similar characteristics into various clusters. This approach will use every data point and store it in a cluster having identical properties. This will be repeated until every data point has been allocated to a cluster. Every generated group will be distinct from the others with varying mean values. This method was chosen because the number of clusters may be fine-tuned before the model is implemented, and the clusters can be limited as desired. Moreover, several related works suggest that the k-means technique outperforms alternative clustering models. To find out the optimum number of clusters, Elbow method and silhouette score is used. Clusters with a specified order from top to bottom are generated using hierarchical clustering. Dendrogram is used to determine the number of clusters for hierarchical clustering. Two clusters are joined in this dendrogram once they are merged, and the height of the join will equal the distance between these locations.

4.2   Customer Lifetime Value Prediction Models

The BG/NBD method is used to determine whether or not the customer is alive/active. It is a prediction system that uses previous customer transactions to identify the number of purchases made by each consumer thereby developing a customer retention strategy. This model performs the prediction of customer purchases that are a portion of customer lifetime value using the recency and frequency scores from the RFM. BetaGeoFitter() function was used and the probability of customers being active is specifically discovered with support of this model.

The Gamma-Gamma method is to predict each customer's monetary value. It calculates the amount spent by each customer using the recency and monetary factors. Lifetime Value (LTV) = future number of transactions * revenue per transaction * margin The output of the BG/NBD model will also be utilized to predict revenue using the R and M parameters in GammaGammaFitter() function.

To report the relationship between two variables, Linear Regression technique fits a linear equation to the observed dataset. The main idea is to create a line that relates the data the best. The line with the minimal overall prediction error is the best line and the error is measured as difference in distance between each data point and regression line.

Random forest is used as it does average to forecast data by fitting a number of classification decision trees on the data sub samples. Due to usage of numerous data instances, overfitting is minimized, and accuracy improved with changes in parameters.

Gradient Boosting is employed along with 3 elements namely a weak learner for making prediction, a loss function which is optimized and additive model to insert the weak learners.

Deep Neural Network are supervised procedures and include three layers namely input, hidden and output layer. Each node is linked to others containing weight and threshold. Node is activated in case the output of node exceeds certain threshold and hence data is passed to next layer. Certain parameters such as epochs, layers, batch size, optimizer, loss, activation function and metrics play a major role.

5   Implementation

The Machine and Deep learning framework was implemented using google Colab and python language was used to carry out this research. The data source was loaded into dataframe using pandas library. The detailed implementation of the code can be viewed in the GitHub repository: https://github.com/kannan-kumaran/Research-Project. Initially, to perform customer segmentation RFM parameters were generated after comprehension of the data. Robust RFM level was measured into 3 categories of customer group such as low, medium and high with consideration of RFM score. For the purpose of comparing the actual RFM levels obtained as discussed above with the ML predicted outcome, clustering techniques were used. K-means identified the optimum number of clusters using elbow method and silhouette score whereas Hierarchical clustering using dendrograms. The plot inferred that the curve flattens after three clusters with minor changes between each cluster. As a result, the customers are segmented in accordance to cluster number (k = 3). Using the Euclidean metric for the linkages, a dendrogram was plotted and based on inference clusters were segregated into 3. Thus K-means algorithm was executed with three clusters and mean of each cluster group was measured. Insights from mean values of recency, frequency and monetary for each cluster appear to be distinct from one another, indicating that customers have been segmented well. Clusters 1 and 2 appear to have higher mean values compared to Cluster 0 which is indication of valuable customers. Customers in Cluster 2 are more valuable than those in Clusters 0 and 1. Similarly, Hierarchical clustering was performed with 3 clusters based on the interpretation of dendograms. Customer lifetime value is a forecast of future buying and spending of each customer. However, in case of machine learning algorithms values have to be trained for prediction as well as need a Y subset that is new to the model for performance evaluation get features() function and data were split into train and test corresponding to dates. The target variable is chosen as the next purchase period and other features as independent variable to perform prediction. For the purpose of identifying the best model, DNN was compared to ML algorithms such as Linear Regression, Random Forest and Gradient Boosting. To forecast future purchases, the target field was chosen as the next purchase period while other features were used as the independent fields. The LinearRegression, Random Forest regressor and GradientBoostingRegressor techniques were imported from sklearn metrics and fitted with training attributes. The actual test variable was compared to predicted test variable by model and evaluated results were stored in dataframe.

6   Evaluation

The evaluation of the Deep Learning Customer Retention framework is performed using two experiments. The first experiment is to evaluate the segmentation of customers in Sect. 6.1. The second experiment is to evaluate the CLV models in Sect. 6.2. The discussions are included in Sect. 6.3.

6.1   Evaluation of Segmentation

The aim of this experiment is to compare customer segmentation undertaken using RFM analysis and clustering methods (K-Means & Hierarchical). To guide retail industry in determination of optimum segmentation, the models had to be compared (Ahalya and Pandey; 2015). This was achieved using heatmap & snake plot for evaluation purpose. The snake plot of K-means clustering for customer profitable classes are shown as (K High, K Medium, K low) respectively in Fig. 3. This plot helps in the classification customers group based on profitability and determines how well are segments created. The peak value represents the high profitable customer class. On inference, the plot produced by K-means indicates more diverse clusters with greater weightage than by manual RFM analysis.

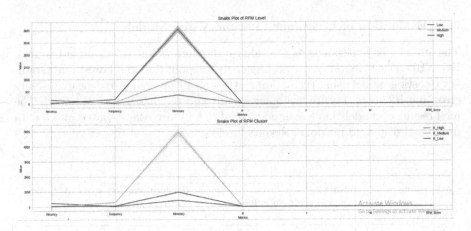

Fig. 3. Snake Plot for comparison of RFM cluster and manual segmentation

Figure 4 demonstrates heat map for RFM segmentation attained with k-means clustering and also with RFM levels. Heat Map is used to find the relative importance and if the segmented groups are widely classified. On comparison of mean values, it was discovered that K-means segmented RFM outperforms manual RFM segmentation level. When comparing high profitable level customers from both heat maps, clustering based segmentation has higher mean values of 2.70 and 2.84 implying that customers are valuable and that segmentation is more effective. The next experiment to evaluate CLV was carried out for the purpose of predicting the customers who will be profitable in future and identify optimum model.

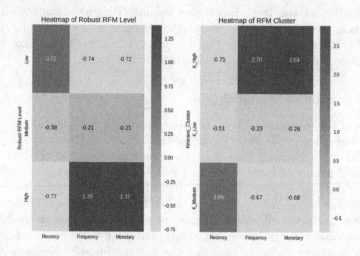

Fig. 4. Heat Map for RFM comparison

6.2   Evaluation of Customer Lifetime Value Models

The aim of this experiment is to evaluate the performance of Gamma-Gamma, BG/NBD, Deep Neural Network, Linear Regression, Random Forest and Gradient Boosting model using evaluation metrics mean squared error, mean absolute error and R2 score. Future customer spends and purchases were obtained. The final results of models implemented to predict CLV are concatenated into a data frame and displayed in Model Results Table.

Model results			
Algorithm	Mean squared error (in million)	Mean absolute error	R^2 score (Accuracy)
Gamma-Gamma	1.8	300	57%
BG/NBD	1.2	0.66	52%
DeepNeuralNetwork	5.9	734	71%
Linear Regression	6.4	760	68%
Random Forest	7.7	811	63%
Gradient Boosting	7.6	840	63%

The BG/NBD model predicted future customer purchases with 51% accuracy. The model accuracy is referred to by the R2 score. The mean absolute error (MAE) was 0.66 when the actual and projected values were compared. This is because models like BG/NBD employ the features period to estimate latent variables. This kind of modeling can be viewed as unsupervised, and our model's features periods are both X and Y.

Then gamma-gamma model was used and evaluated for customer spend forecast that produces 57% accuracy. The error values are MAE - 300. This model did not perform well in forecasting future spending.

After application of probabilistic models, Machine Learning algorithms were implemented. Linear Regression resulted with R2 score-68%, mean absolute error-760. Random Forest resulted with R2 score-63%, mean absolute error-811. Gradient Boosting resulted with R2 score-63.4%, mean absolute error-840. In order to identify whether it is feasible to obtain an improved model accuracy by Deep Learning, DNN was implemented and outperformed the rest of the models by generating results with the highest accuracy of 71% and lower mean absolute error value-734.

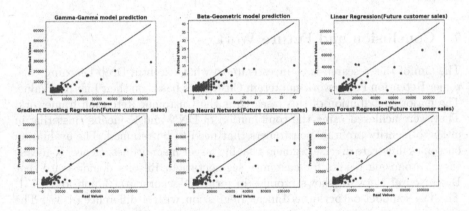

Fig. 5. Actual vs Predict CLV model results

The comparison of actual and forecasted values for all five models are shown in Fig. 5 and notice that few models have forecasted values close to actual values. Thus, guides in better model identification through visualizations. However, taking into account both model results table and plots, shows promise for DNN with best accuracy and contributes a higher performance than beta-geometric, gamma-gamma model, Linear Regression, Random Forest and Gradient Boosting Regression.

6.3 Discussion

With an emphasis on CLV adoption, the suggested Machine and Deep Learning framework for customer retention was created. As a result, the retail firm sales were examined using historical transactions and CLV was projected using multiple models to assure high accuracy, increased performance, and little loss. The use of Deep Learning in this study distinguishes it from past relevant work. The comparison of DNN results with probabilistic and ML models confirmed the need of developing a better performance model for accurate predictions. Furthermore, ML clustering integration is an extra benefit to strategical marketing for classifying profitable classes such as high, medium, and low level. The inferences from exploratory data analysis are customers that has placed the most orders are from the United Kingdom and that spends the most price on purchases is from the Netherlands. Product in demand was "White Hanging T-light Holder". The month of November in year 2011 had the highest sales. According to days of week, Monday to Thursday faced a steady increase and later decreases. Deep Neural Network shows promise if the motivation is for accuracy as well as loss. RFM values obtained by clustering are promising with more diverse clusters and weightage than manual RFM calculation. Improved performance and accuracy is possible with the availability of additional data points. A minimum of 3 years of transaction history could achieve more data which is thereby a limitation in this research.

7   Conclusion and Future Work

The aim of this research is to investigate to what extent a Deep Learning Customer Retention Framework can retain Customers based on their Lifetime Value. The major contribution of this research is the Customer Retention framework. They were achieved using the combination of DNN and K-means clustering in order to identify profitable customers that need to be retained. The evaluation outcomes illustrate diverse segments having larger mean values. Hence, K-means shows promising results for customer segmentation. Results demonstrate that Deep Neural Network shows promise in terms of accuracy-71% and loss - 731. The DNN model outperforms Linear Regression with a difference of 3%. The future work could be to optimize by experimenting new features and data samples for feasible model training. To improve the accuracy of the used models, a larger sample size and a longer time span can be utilized as input. Also, hyperparameter tuning on the parameters and integration of algorithms can be carried out as part of future study.

References

1. Chang, W., Chang, C., Li, Q.: Customer lifetime value: a review. Soc. Behav. Personal. Int. J. **40**, 1057–1064 (2012)
2. Win, T.T., Bo, K.S.: Predicting customer class using customer lifetime value with Random Forest algorithm. In: 2020 International Conference on Advanced Information Technologies (ICAIT) (2020)
3. Monalisa, S., Nadya, P., Novita, R.: Analysis for customer lifetime value categorization with RFM model. Procedia Comput. Sci. **161**, 834–840 (2019)

4. Venkatakrishna, M.R., Mishra, M.P., Tiwari, M.S.P.: Customer lifetime value prediction and segmentation using machine learning (2021)
5. Singh, L., Kaur, N., Chetty, G.: Customer life time value model framework using gradient boost trees with RANSAC response regularization. In: 2018 International Joint Conference on Neural Networks (IJCNN) (2018)
6. Wu, J., et al.: An empirical study on customer segmentation by purchase behaviors using a RFM model and K-means algorithm. Math. Probl. Eng. **2020**, 1–7 (2020)
7. Kansal, T., Bahuguna, S., Singh, V., Choudhury, T.: Customer segmentation using K-means clustering. In: 2018 International Conference on Computational Techniques, Electronics and Mechanical Systems (CTEMS), Conference on Neural Networks (IJCNN) (2018)
8. Chen, D.: A real online retail transaction data set of two years, 2009–2011 [Data set]. https://archive.ics.uci.edu/ml/datasets/Online+Retail+II

A Deep Learning Based Approach to Automate Clinical Coding of Electronic Health Records

Ashutosh Kumar and Santosh Singh Rathore[✉][iD]

Department of Computer Science and Engineering, ABV-Indian Institute of Information Technology and Management, Gwalior, India
{imt_2017022,santoshs}@iiitm.ac.in

Abstract. The medical records in different electronic formats, such as handwritten notes, diagnosis summaries, lab reports, electronic pdfs, etc., contain valuable information that can be used for various medical purposes. These health records are currently coded manually or semi-automated to assign clinical codes (ICD-codes) for clinical research and analytics. This process is very time-consuming, expensive, and error-prone. This paper presents a method for automated clinical coding of electronic health records (EHRs) given the patient diagnosis summary and other medical-related documents. The presented method uses natural language processing (NLP) techniques, which capture knowledge from the free-text diagnosis descriptions, do the text matching and semantic mapping, and translate diagnosis descriptions into clinical codes. We develop one baseline *Word2vec* and cosine similarity hybrid model, a transformer encoder model, and a BERT (Bidirectional Encoder Representations from Transformers) model for the automated clinical coding. The presented models are evaluated using a publicly available Medical Information Mart for Intensive Care III (MIMIC-III) dataset. The used dataset consists of various patient diagnosis descriptions and corresponding ICD-9 codes. The experimental results show that the presented *BlueBERT* based automated clinical coding model produced an AUC (area under ROC curve) value of 98.9% for the top-10 ICD codes prediction. On the full MIMIC-III dataset, the transformer model produced an accuracy of 76.8%, a precision of 61.02%, a recall of 47.22%, a f1-score of 53.2%, and an AUC value of 92.1%. The hybrid baseline model and another used transformer encoder model also showed promising results.

Keywords: Machine learning · Electronic health record · Clinical coding · Cosine similarity · Transformers

1 Introduction

Nowadays, patient hospital visits and medical records such as handwritten notes, printed prescriptions, discharge summaries, etc., are recorded in electronic format [1]. These records are known as Electronic Health Records (EHRs). The

P. P. Roy et al. (Eds.): BDA 2022 India, LNCS 13773, pp. 104–116, 2022.
https://doi.org/10.1007/978-3-031-24094-2_7

EHRs are assigned international Classification of Diseases (ICD) codes for making clinical decisions and insurance and billing purposes. This task is done manually, which is highly time consuming, error prone and financially expensive [2]. Furthermore, it is prone to errors due to human-driven activity. A system that could perform ICD coding automatically can be a very useful tool, and it can save a great amount of resources and speed up the process [3].

An automated clinical coding system can analyze EHRs and accurately determine the relevant clinical codes using AI-based approaches. Generally, these automated systems aggregate the medical data of a patient available in different formats and ingest and clean the data. Subsequently, the data is processed by identifying relevant medical terminologies and doing entity recognition. Finally, some learning techniques (machine learning or deep learning) are applied to predict appropriate ICD codes to the patient's medical records [4–6]. Later, these codes are validated by the hospital's coding team. Previously, many researchers have used different machine learning techniques such as logistic regression, K-nearest neighbors, naive Bayes, support vector machines, etc. to build automated clinical coding systems. These techniques use ICD code data stores in the form of EHRs with the corresponding ICD code label [7–9].

The complexity of medical language is the biggest barrier to developing automated clinical coding systems. The medical records are generally available in the free-text form, and contain unclear vocabulary, non-standard syntax, and ambiguous abbreviations. Furthermore, the records contained many unpopular words, slang, spelling, and grammatical errors [10]. All these problems collectively led to the poor performance of the earlier developed automated clinical coding systems. Recently, many sophisticated natural language processing (NLP) techniques have been developed, extracting relevant information from the free text and further using this information to build prediction models. These techniques can automatically extract the language-related features, thus removing the need for manual feature engineering. Motivated by this, in this paper, we explore the use of state-of-the-art NLP techniques to build an automated clinical coding system. Specifically, we develop one baseline *Word2vec* and cosine similarity hybrid model, a transformer encoder model, and a BERT model for the automated clinical coding. The presented models are trained and tested on the Medical Information Mart for Intensive Care III (MIMIC-III) dataset, which is available in the public domain. The dataset contained different EHRs of several patients and corresponding ICD-9 codes. The results are reported in terms of accuracy, precision, recall, and AUC measures. The experimental results showed that the presented BERT-based model (BlueBERT) yielded the best performance than other models. All the developed models produced state-of-the-art performance.

The rest of the paper is organized as follows. Section 2 presents related work on ICD-code prediction. Section 3 describes the presented method for ICD code prediction. Experimental setup and analysis are presented in Sect. 4. Conclusions follow it in Sect. 5.

2 Related Work

With the progress of machine learning over the recent years, a significant progress has been observed in efforts to solve the problem of automating the clinical coding through classifier models. This section discusses some of the recent works related to the automated ICD-code prediction.

Catling et al. [10] presented an approach toward automated clinical coding. To build the model, the authors used a hierarchical clinical coding ontology and TF-IDF for the text representation. The model was validated on the MIMIC-III dataset for the ICD-9 code prediction. The results showed that the model achieved an F1-score of 0.682–0.701 for the prediction of the 19 disease-category labels. In another work, Sonabend et al. [11] reported a model for automated ICD coding using unsupervised knowledge integration (UNITE). The model was validated using coded ICD data for 6 diseases from PHS and MIMIC-III datasets. The results showed that the presented UNITE model achieved an average AUC of 0.91 at PHS and 0.92 at MIMIC over 6 diseases. Diao et al. [12] used a clinically interpretable model for the automated ICD coding for primary diagnosis. The model predicts ICD-10 codes from 168 primary diagnoses. The model produced accuracy and macro averaged F1 (Macro-F1) of 95.2% and 88.3%, respectively.

Purushotham et al. [13] performed a benchmark analysis of deep learning techniques for the MIMIC-III dataset. The authors used different deep learning models, ensemble models, and super learner algorithms and found that deep learning models consistently outperform all the other used models. In another similar work, Shi et al. [14] explored deep learning models for automated ICD clinical coding. The authors developed a hierarchical deep learning model with an attention mechanism for the prediction. The presented model achieved 0.53 and 0.90 F1-score and AUC values, respectively. Dong et al. [15] proposed an explainable model for clinical notes using hierarchical label-wise attention networks and label embedding. The model was applied to predict top-50 ICD-codes, and results showed that the model produced the best AUC and F1 on the MIMIC-III dataset, 91.9%, and 64.1%, respectively.

Perotte et al. [16] used SVMs and treated each ICD-9 code as an independent label. It resulted in a huge binary classification problem, considering each prediction output as a separate result. The result was thus called a flat classifier and showed average results. Zhang et al. [17] and Ayyar et al. [18] summarized the research on clinical coding and used advanced neural architecture such as LSTM. It is generally used for sequence encoding and language modeling. It helped in defining the long-distance relation between the tokens in clinical notes. Mullenbach et al. [19] used base machine learning models such as logistic regression and treated the prediction as a multi-label text classification problem. They used a convolution-based approach and an attention model, which achieved the current state-of-the-art results with an F1 score of 0.53 on MIMIC-III Diagnosis codes. Moons et al. [20] compared the performance of recent base convolutional and recurrent models such as BiGRU (Gated Recurrent Unit) with more advanced clinical coding approaches like DR-CAML, MVC-LDA, and MVC-RLDA. They provided a survey of recent neural approaches for ICD-9 coding on the MIMIC-

III dataset. Diao et al. [21] used a baseline machine learning model with one-hot encoding on Chinese EHRs and a Light Gradient Boosting Machine (LightGBM) classifier for multi-label text classification problems. The authors achieved an F1 score of 0.70 in the Chinese context for automated clinical coding. Moqurrab et al. [22] developed a deep learning based model for the entity recognition from clinical notes. The presented model was a combination of CNN, Bi-LSTM, and CRF with non-complex embedding. The results showed that the presented model outperformed existing models by a margin of 4–10% and 5–12% in terms of F1-score on i2b2-2010 and i2b2-2012 data.

3 Presented Automated Clinical Coding Models

3.1 ICD-9 Codes

We have used the ICD-9 medical coding scheme for the clinical coding of electronic health records [23]. These medical codes have at most five numbers based on hierarchy taxonomy (Shown in Fig. 1). The pattern is quite simple to understand. The first three numbers of the code are used to represent the top level of a particular disease. The next number marks the specificity of the disease, and the fifth number is used to categorize different variants of a particularly high-level disease. The figure shows an ICD-9 coding example.

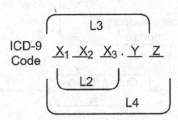

240-279: Endocrine, nutritional and metabolic diseases   (L1)
250:      Diabetes mellitus   (L2)
250.2:   Diabetes with hyperosmolarity   (L3)
250.21:   Diabetes with hyperosmolarity of type 1   (L4)

Fig. 1. ICD-9 code structure with second through fourth layer representations and diabetes as an example [20]

3.2 Presented Models

An overview of the methodology used to design the automated clinical coding system is shown in Fig. 2. The step is acquiring the EHRs dataset. We use two kinds of health records presented in the MIMIC-III dataset[1]: complete health notes from the NOTEEVENTS table and discharge summaries.

---

[1] https://physionet.org/content/mimiciii-demo/1.4/.

The NOTEEVENTS table contains clinical free-text notes, including nursing and physician notes, ECG (electrocardiogram) reports, and radiology reports. Discharge summaries include the reason for hospitalization, primary diagnosis, procedures and treatment, and the patient's discharge condition. In this work, two different categories of models are used. A baseline model is based on a dictionary-based hybrid word2vec and cosine similarity module. The other models are the transformer ones. We use two different types of transformer models, a simple transformer encoder model and a BlueBERT model that we fine-tune for the discharge summaries information. The output of all these models is the respective ICD-9 codes corresponding to the respective medical terms.

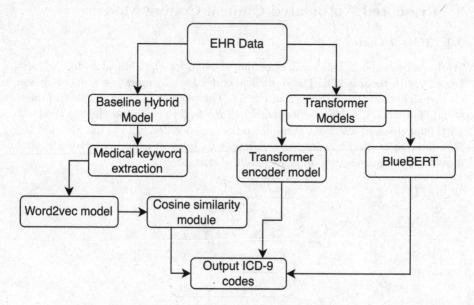

Fig. 2. Overview of the presented methodology

3.3   Baseline Word2vec and Cosine Similarity Hybrid Model

Figure 3 provides an overview of the presented baseline model. The model is a hybrid of the word2vec embedding and cosine similarity. The model is designed as follows. The input to the model is EHR free-text dataset. The first step extracts the medical keywords from the input data. We used the keyword extraction module using Zhang et al. [17] for this. In the second step, NLP pre-processing is applied. It includes vectorization and the word2vec technique, which map words or phrases from vocabulary to a corresponding vector of real numbers and extract the notion of relatedness across words in the text. Thereafter, a cosine similarity module is applied to measure the text similarity between two documents. The generated vectors after the second step are compared to the NOTEEVENTS data. Finally, an appropriate ICD-9 code is assigned to the input EHR text. The whole pipeline works on a clinical text document as an input, and the output is a list of medical keywords coded per the ICD-9 coding system.

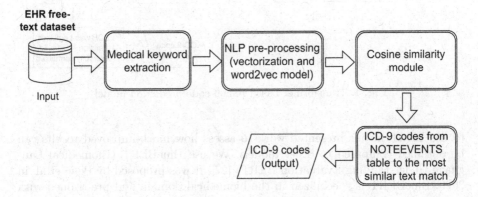

Fig. 3. Baseline hybrid ICD-9 code prediction model

3.4 Transformer Encoder Model Results

Figure 4 provides description of the presented transformer based ICD-9 code prediction model. The working of the model is as follows. The model takes EHR free-text data as clinical notes as the input. In the first step, an embedding layer is applied. This layer indexes each word of the record and converts it to a set of word embedding vectors. The second step is the use of a transformer encoder. This layer has a multi-head self-attention method for sequence representation. It helps in representing the vectors contextually. The layer has a set of 6 encoding layers that have eight attention heads with 256 features. At this stage, a special loss function converts a simple text-classification problem to a multi-label classification problem. BCEWithLogits loss function is used with Mean Reduction [24]. Finally, every output was passed from the previous stage to a fully connected Sigmoid function and predicted the ICD-9 code per pre-set threshold values.

3.5 BERT Model (BlueBERT)

Additionally, we explore using the BERT model for the ICD-9 code prediction of the given EHR data. Bidirectional Encoder Representations from Transformers (BERT) is a state-of-the-art model that is based on the multi-layer bidirectional transformer architecture. It uses a feed-forward multi-headed self-attention encoder that is trained with two objectives, MLM (masked language modeling basically predicts a missing word in a sentence from the context) and NSP (next sentence prediction predicts whether two sentences follow sentences). Many NLP-related works showed that BERT had improved the performance in language comprehension and processing [25,26]. Therefore, it was decided

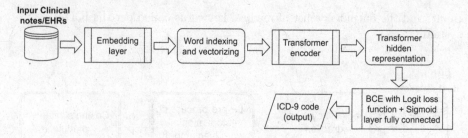

Fig. 4. Transformer based ICD-9 code prediction model

to use BERT in our presented work to assess how much-improved results can be obtained compared to other models. We use BlueBERT (Biomedical Language Understanding Evaluation BERT) [27]. It was proposed by Peng et al. in 2020. BlueBERT is specialized in the biomedical domain and pre-trained with and fine-tuned for five tasks (Sentence Similarity, Relation Extraction, Named Entity Recognition NER, Document Classification, and Inference). The details of this model can be referred to from [27].

4   Experimental Analysis and Results

4.1   Used MIMIC-III Dataset

In the presented work, the MIMIC-III (Medical Information Mart for Intensive Care III) dataset is used to build and evaluate the presented ICD code prediction models [28]. It is a large open-source dataset containing de-identified medical data of over 40,000 patients staying in the Beth Israel Deaconess Medical Center between 2001 and 2012. The dataset contains demographical information, vital signs, lab tests, procedures, medications, mortality outcomes, and clinical notes. The primary focus was on the free-text clinical notes written by physicians describing their patients in the intensive care unit (ICU). The dataset is vast, and there are 26 tables of de-identified patient data. We have performed exploratory data analysis and the pre-processing of the dataset at the same stage. It was observed that the MIMIC-III dataset has 52,726 discharge summaries, which in total have 8,929 unique ICD-9 codes. These codes are further partitioned into two categories, namely Diagnosis and Procedure. The top-10 most frequently occurring ICD-9 codes in the dataset were also interpreted (Shown in Fig. 5). The ICD-9 codes were specifically used because of two main reasons. The first reason is the easy accessibility and huge data collection in MIMIC-III from 2001 to 2012. The second reason is that all the previous benchmarks specifically perform on ICD-9 codes. We performed the basic NLP pre-processing on the discharge notes in the NOTEVENETS table of the MIMIC-III dataset.

– Lowercased all the words.
– Tokenized the text into words.

- Removed all punctuation, numbers and special characters.
- All the English stopwords were removed.
- The final set of tokens were stemmed using the Snowball stemming.

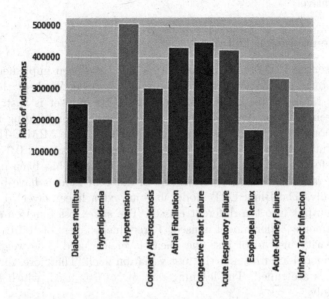

Fig. 5. Top-10 most frequent ICD-9 codes in dataset

4.2 Evaluation Metrics

For the evaluation purpose, five different metrics, accuracy, precision, recall, ROC curve (AUC), and F1 score are used. The AUC - ROC curve measures the performance of classification problems at various threshold settings. ROC is a probability curve while the AUC represents the degree or measure of separability.

Accuracy is the fraction of the examples that are correctly predicted. It is defined by Eq. 1. Precision is the proportion of the positive identified labels that are actually correct. It is defined by Eq. 2. Recall is the proportion of actual positives that are identified correctly. It is defined by Eq. 3.

$$Accuracy = \frac{TP + TN}{TP + TN + FN + FP} \tag{1}$$

$$Precision = \frac{TP}{TP + FP} \tag{2}$$

$$Recall = \frac{TP}{TP + FN} \tag{3}$$

F1-score which is defined as the harmonic mean of the model's precision and recall is given in Eq. 4.

$$F1 - score = \frac{2 * Precision * Recall}{Precision + Recall} \tag{4}$$

where, TP = True positive, FP = False positive, and FN = False negative, TN = True negative.

4.3 Implementation Details

All the presented ICD-9 code prediction models have been implemented using different libraries of the Python programming language. We kept the training and testing ratio as 90:10. Because the MIMIC-III dataset is vast, this data split has been used for the experiments. The Word2vec embedding model has 100 dimensions layer, which is trained specifically for the MIMIC-III vocabulary. A learning rate of 1e−4 and the dropout probability was 0.3 were used in the transformer models. BCEWithLogits loss function has been used in the transformer model. The other parameters of different models have been tuned experimentally. The BlueBERT model uses an extra linear layer to transform the model output into ten different classes. The same loss function as used in the transformer encoder model is used. Three linear layers of ReLU activation for non-linearity instead of just one linear layer are used. The weights of the BlueBERT model are frozen at the first variation itself. Therefore, further, only linear weights are tuned. The learning rate of 2e−5 is used, which is selected experimentally.

4.4 Results and Analysis

Table 1 shows the results of different presented models in terms of accuracy, precision, recall, f1-score, and AUC measures. The reported results are only for top-10 ICD-9 codes available in the dataset. From the table, it can be seen that all four presented models achieved an AUC value of more than 90%, with the highest value of 98.9% by the BlueBERT model. On the full MIMIC-III dataset, the transformer model produced an accuracy of 76.8%, a precision of 61.02%, a recall of 47.22%, a f1-score of 53.2%, and an AUC value of 92.1%. Except for the recall measure, all other values are improved compared to the presented baseline model. BlueBERT is the best performing model among all the used ones.

Confidence Interval Analysis of the Presented Baseline Model. The presented baseline hybrid model is evaluated for different confidence interval (CI) levels based on the cosine similarity. Table 2 and Fig. 6 depict the results of this analysis for AUC and PR curve measures. The CI level has varied from 100% to 60%, and results are recorded. It can be observed that the AUC value varied from 93.8% (max) to 76.8% (min). These results showed that the presented baseline model produced improved results for the ICD-9 code prediction.

Table 1. Results of the presented ICD-9 code prediction models

Models	Accuracy	Precision	Recall	F1-score	AUC (micro)
Word2vec + Cosine similarity model (Data: All MIMIC III clinical notes)	74.69%	51.33%	64.21%	57.05%	92.1%
Transformer encoder model (Data: All MIMIC III clinical notes)	76.84%	61.02%	47.22%	53.24%	92.1%
Transformer encoder model (Data: Only MIMIC III discharge summary)	80.58%	76.63%	71.61%	74.03%	95.4%
BERT model: BlueBERT (Data: Only MIMIC III discharge summary)	84.49%	78.60%	63.62%	70.2%	98.9%

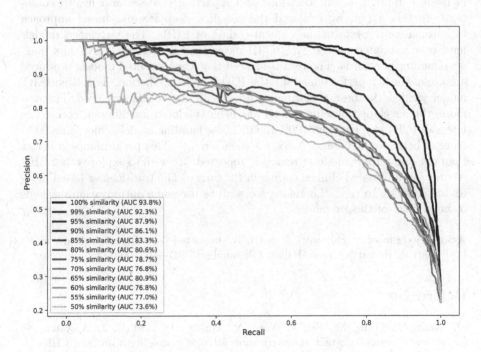

Fig. 6. PR curve for baseline hybrid model

Table 2. AUC scores with different confidence intervals for Baseline hybrid model

Confidence Interval (based on similarity scores)	AUC score (in %)
100%	93.8
99%	92.3
95%	87.9
90%	86.1
85%	83.3
80%	80.6
75%	78.7
70%	76.8
65%	80.9
60%	76.8

5 Conclusion

Automated clinical coding of the EHR records has many benefits and can assist in medical reimbursement decisions and reporting diseases and health conditions. In this paper, we explored the use of a deep learning-based approach to automatically perform the clinical coding of EHRs. The developed models have been validated on the MIMIC-III dataset using NOTEEVENT and diagnosis summary records. The results showed that the presented models produced a state-of-the-art performance for the ICD-9 code prediction. The BlueBERT model yielded the best performance among all the techniques used. Furthermore, the developed baseline model has been validated for different confidence intervals (CIs) ranging from 100% to 60%. The baseline model achieved an AUC score of between 93.8% and 76.8% for the CI range. This performance is better than the counterpart models previously reported. We wish to explore other EHR records for automated clinical coding in the future. The transformer-based models are complex to tune. Therefore, we wish to use some optimization methods to build state-of-the-art models.

Acknowledgement. This work is partially supported by a Research Grant under DST-Start up Research Grant (India), File number: SRG/2021/000173.

References

1. Wang, M., Wang, M., Fei, Y., Yang, Y., Walker, J., Mostafa, J.: A systematic review of automatic text summarization for biomedical literature and EHRs. J. Am. Med. Inform. Assoc. **28**(10), 2287–2297 (2021)
2. Subotin, M., Davis, A.: A system for predicting ICD-10-pcs codes from electronic health records. In: 2014 Proceedings of BioNLP, pp. 59–67 (2014)

3. J., Teng, F., Ma, Z., Chen, L., Huang, L., Li, X.: A multi-channel convolutional neural network for ICD coding. In: 2019 IEEE 14th International Conference on Intelligent Systems and Knowledge Engineering (ISKE), pp. 1178–1184. IEEE (2019)

4. Zhang, Z., Liu, J., Razavian, N.: BERT-XML: large scale automated ICD coding using BERT pretraining. arXiv preprint arXiv:2006.03685 (2020)

5. Xie, X., Xiong, Y., Yu, P.S., Zhu, Y.: EHR coding with multi-scale feature attention and structured knowledge graph propagation. In: Proceedings of the 28th ACM International Conference on Information and Knowledge Management, pp. 649–658 (2019)

6. Rubbo, B., et al.: Use of electronic health records to ascertain, validate and phenotype acute myocardial infarction: a systematic review and recommendations. Int. J. Cardiol. **187**, 705–711 (2015)

7. Atutxa, A., Pérez, A., Casillas, A.: Machine learning approaches on diagnostic term encoding with the ICD for clinical documentation. IEEE J. Biomed. Health Inform. **22**(4), 1323–1329 (2017)

8. Xu, K., et al.: Multimodal machine learning for automated ICD coding. In: Machine Learning for Healthcare Conference, pp. 197–215. PMLR (2019)

9. Jamian, L., Wheless, L., Crofford, L.J., Barnado, A.: Rule-based and machine learning algorithms identify patients with systemic sclerosis accurately in the electronic health record. Arthritis Res. Therp. **21**(1), 1–9 (2019)

10. Catling, F., Spithourakis, G.P., Riedel, S.: Towards automated clinical coding. Int. J. Med. Inform. **120**, 50–61 (2018)

11. Sonabend, A., et al.: Automated ICD coding via unsupervised knowledge integration (unite). Int. J. Med. Inform. **139**, 104135 (2020)

12. Diao, X., et al.: Automated ICD coding for primary diagnosis via clinically interpretable machine learning. Int. J. Med. Inform. **153**, 104543 (2021)

13. Purushotham, S., Meng, C., Che, Z., Liu, Y.: Benchmarking deep learning models on large healthcare datasets. J. Biomed. Inform. **83**, 112–134 (2018)

14. Shi, H., Xie, P., Hu, Z., Zhang, M., Xing, E.P.: Towards automated ICD coding using deep learning. arXiv preprint arXiv:1711.04075 (2017)

15. Dong, H., Suárez-Paniagua, V., Whiteley, W., Honghan, W.: Explainable automated coding of clinical notes using hierarchical label-wise attention networks and label embedding initialisation. J. Biomed. Inform. **116**, 103728 (2021)

16. Perotte, A., Pivovarov, R., Natarajan, K., Weiskopf, N., Wood, F., Elhadad, N.: Diagnosis code assignment: models and evaluation metrics. J. Am. Med. Inform. Assoc. **21**(2), 231–237 (2014)

17. Zhang, Y., Zhang, Y., Qi, P., Manning, C.D., Langlotz, C.P.: Biomedical and clinical English model packages for the stanza python NLP library. J. Am. Med. Inform. Assoc. **28**(9), 1892–1899 (2021)

18. Ayyar, S., Don, O., Iv, W.: Tagging patient notes with ICD-9 codes. In: Proceedings of the 29th Conference on Neural Information Processing Systems, pp. 1–8 (2016)

19. Mullenbach, J., Wiegreffe, S., Duke, J., Sun, J., Eisenstein, J.: Explainable prediction of medical codes from clinical text. arXiv preprint arXiv:1802.05695 (2018)

20. Moons, E., Khanna, A., Akkasi, A., Moens, M.-F.: A comparison of deep learning methods for ICD coding of clinical records. Appl. Sci. **10**(15), 5262 (2020)

21. Jiang, Z., et al.: A light gradient boosting machine-enabled early prediction of cardiotoxicity for breast cancer patients. Int. J. Radiat. Oncol. Biol. Phys. **111**(3), e223 (2021)

22. Moqurrab, S.A., Ayub, U., Anjum, A., Asghar, S., Srivastava, G.: An accurate deep learning model for clinical entity recognition from clinical notes. IEEE J. Biomed. Health Inform. **25**(10), 3804–3811 (2021)

23. Wei, M.Y., Luster, J.E., Chan, C.-L., Min, L.: Comprehensive review of ICD-9 code accuracies to measure multimorbidity in administrative data. BMC Health Serv. Res. **20**(1), 1–11 (2020)

24. Zhang, Y., Lu, Z., Wang, S.: Unsupervised feature selection via transformed auto-encoder. Knowl.-Based Syst. **215**, 106748 (2021)

25. Tenney, I., Das, D., Pavlick, E.: BERT rediscovers the classical NLP pipeline. arXiv preprint arXiv:1905.05950 (2019)

26. Acheampong, F.A., Nunoo-Mensah, H., Chen, W.: Transformer models for text-based emotion detection: a review of BERT-based approaches. Artif. Intell. Rev. **54**(8), 5789–5829 (2021)

27. Peng, Y., Chen, Q., Lu, Z.: An empirical study of multi-task learning on BERT for biomedical text mining. arXiv preprint arXiv:2005.02799 (2020)

28. Johnson, A.E.W., et al.: MIMIC-III, a freely accessible critical care database. Sci. Data **3**(1), 1–9 (2016)

Determining the Severity of Dementia Using Ensemble Learning

Shruti Srivatsan[1]([✉]), Sumneet Kaur Bamrah[1], and K. S. Gayathri[2]

[1] Department of Computer Science and Engineering, Sri Venkateswara College of Engineering, Sriperumbudur, Tamil Nadu, India
shrutisrisvce@gmail.com
[2] Department of Information Technology, Sri Sivasubramaniya Nadar College of Engineering, Kalavakkam, Tamil Nadu, India
gayathriks@ssn.edu.in

Abstract. According to WHO, 10 million individuals are subject to the risk of dementia annually. Geriatric care is expensive, restricted, and inaccessible for dementia care. The condition has no cure and medication is provided to inhibit the progress. Individuals exhibit a decline in cognition at severe stages impacting behaviors, hence detection of the early onset of dementia is critical. As an individual ages, the symptoms of dementia become harder to distinguish from symptoms of normal ageing. Behavioral patterns are identified using Activities of Daily Living (ADL) data. Traditional forms of dementia prediction widely use Magnetic Resonance Imaging (MRI). The detection of dementia and the severity utilizes both forms of data indvidually largely. A novel approach is proposed to combine motion sensor and brain scan data alternatively to detect the presence and eventually severity of dementia. Patterns of behavior are recorded, the interactions between the occupant and sensors placed in a smart environment. Daily activity recordings are used to detect the presence or absence of dementia in Phase 1. Phase 2 focuses on the utility of MRI data in longitudinal and cross-sectional form to further assess the severity. Distinct features are extracted using feature selection method. The hyperparameters are tuned and stratified k-fold is applied as well. In both phases, Random Forest classifier performs effectively generating an accuracies of 95.74% and 90.29% in Phase 1 and Phase 2 respectively. Dementia prediction and severity prediction can be extended to provide direct support to members of the assitive care community.

Keywords: Dementia · Activities of daily living · Magnetic resonance imaging · Machine learning · Ensemble learning · Smart healthcare

1 Introduction

Dementia is closely associated with cognition. When two or more functions of cognition showcase deterioration, a clinical state emerges, known as dementia. Nerve cells or neurons, used in cognition get damaged or altered due to chronic

© The Author(s), under exclusive license to Springer Nature Switzerland AG 2022
P. P. Roy et al. (Eds.): BDA 2022 India, LNCS 13773, pp. 117–135, 2022.
https://doi.org/10.1007/978-3-031-24094-2_8

conditions. The brain comprises an outer layer or cerebral cortex made up of gray matter and is responsible for higher-order functions. The region below the cortex or subcortex consists of white matter and is responsible for primitive functions. Based on where lesions are formed or detected, dementia is categorized accordingly. When vascular lesions occur in the subcortical region, subcortical dementia is identified [29] Neurons shrink when an individual is affected by the disorder which can be observed as seen in Fig. 1.

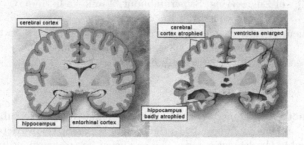

Fig. 1. Presence of cortical atrophy observed in dementia (National Institute on Aging)

The most common forms of dementia are Alzheimer's Disease (AD), Vascular Dementia (VD), Parkinson's Disease (PD), and Frontotemporal dementia (FTD) as indicated in Table 1. Rare and distinct forms of dementia include Creutzfeldt-Jakob disease (CJD), Huntington's, special cases of multiple sclerosis (MS), amyotrophic lateral sclerosis (ALS), leukoencephalopathies, and multiple system atrophy [2].

Table 1. Broad classification of dementia based on the section of the brain impacted

Factors	Cortical Dementia	Subcortical Dementia
Section of the brain impacted	Gray matter (in the cortex)	White matter (below the cortex)
Key elements for identification	Irregularity in behaviors Loss of memory Unable to use language normally	Abnormal brain functioning Decline in emotional capacity Language, memory function intact
Example diseases	Alzheimer's Disease (AD) Creutzfeldt-Jakob Disease (CJD) Frontotemporal dementia (FTD)	Parkinson's Disease (PD) Huntington's Disease Vascular Dementia (VD)

The overall health and lifestyle of the individual are impacted. The world's leading organizations take consistent and persistent efforts to share and disseminate relevant knowledge about the illness [1, 2, 5, 6]. Older people are subject to

risks of memory loss and the number of dementia cases is on a steady rise according to various statistics and reports [4]. The WHO has estimated the number of
patients succumbing to dementia would be 82 million in 2030 and 152 million in
2050. The global social care and medical costs of dementia were reported to be
818 billion dollars in 2015. High-income countries are able to provide superior
quality of care and assistance to dementia patients in comparison to lower or
lower-middle-income countries since expert medical care is unevenly distributed
or even inaccessible. Economically developing countries feel the burden in the
long term. Geriatrics is a specialized field dedicated to addressing the healthcare
needs of elderly individuals. Sophisticated equipment and devices are used to
capture and monitor the regular routine activities to notice any inconsistencies.

Researchers explore various methods to diagnose the onset and the different stages of dementia. If there are noticeable significant changes in performing
daily routine activities, the presence of dementia is captured [17]. Tests are
performed at regular intervals over prolonged periods of time to estimate the
same accurately. The information is further used to clearly identify the relative
stages of progress in the illness is detected. Since routine activities are impacted,
behavioral patterns are gradually influenced as well [24]. Capturing the variations continually requires medical intervention. Detection of dementia and its
severity prediction is a challenging and critical phase for the individuals and
clinicians involved. The methodic capture of ADL data is an alternate option
being explored in dementia detection. Occupational, physical, and speech thera-
pists capture and assess other specialized forms of data in smart environments.
There are various tools used to examine, measure and analyze data. Information
is captured using accelerometers, motion detectors, and cameras. Wearable and
non-wearable sensors are used to detect the movements of the individuals monitored. Functional Independence Measure (FIM), Barthel Index (BI), Katz Index
of Independence are examples of assessment tools used to detect problems related
to ADL. Existing studies indicate the use of different forms of data utilized in
varied scenarios [14,30]. Detection of dementia using neuroimages is a long and
labor-intensive process given the magnitude of factors taken into account for the
process. Specialized and sophisticated machines are required to capture the necessary details over a series of interactions. The individuals are elderly patients
who need extra support and care during the diagnosis process. Since the syndrome is progressive in nature, neuroimages can be used as a final measure to
detect the severity of the syndrome in addition to several other alternatives that
can also capture behavioral patterns and relevant essential details.

Symptoms of dementia at an early stage and in progression can be difficult
to distinguish for clinicians and radiologists. Specialized scans such as PET,
SPECT, and CT are effectively used to diagnose Mild Cognitive Impairment
(MCI) and enable dementia detection. In case of the disease progression, a thorough and detailed record of the patient's activities is maintained. A critical
element in the diagnosis process is the age where the cognitive activities are
assessed over an elongated period of time. Questions and answers are recorded
in the interactive sessions to better understand the individual's cognitive skills.

In addition, physical examinations and a mental status test are performed for further accumulation of relevant details in the diagnosis process.

The number of patients who have dementia is impacted socially, financially, and culturally, thus there is an immediate need to diagnose individuals with associated symptoms. It becomes crucial to explore and formulate plans to take relevant preventive measures to assist the individual impacted. Applications of artificial intelligence are widely soaring in the current generation. Machine learning and deep learning algorithms are developed for specific scenarios to assist the medical community. The benefit is further offered to the individuals seeking medical attention. Drug discoveries are being made to inhibit the progress of dementia [10]. Machine Learning offers additional benefits of working effectively with numerical forms of data. In the work presented, a classification mechanism is implemented in two phases. The first phase focuses on the detection of dementia with ADL data and the second phase is on predicting the severity using specialized forms of data. The objective is to be able to build a strong and robust framework that reduces the timeline for detecting the illness with a combination of approaches ensuring novelty.

In order to determine the early stages of dementia, individuals are carefully monitored for several weeks or even months. The process is slow and tedious, as multiple rounds of tests and scans are taken for patients over consistent sessions, some even spanning years. The use of both image and numerical data promotes better prospects of detecting and classifying if the patient has dementia or not. In this paper, a unique approach has been used by developing a framework that combines different forms of data in a multi-phase approach to address the issue [31].

Some of the key highlights of the study include,

- ADL data is used distinctly to identify the onset of dementia without the use of commonly used neurological biomarkers. It disables the approach of using a scan at the initial stage.
- Use of combinational MRI data is explored. Unique and relevant features are combined using feature selection for the prediction of the severity of dementia.
- Machine learning algorithms are preferred with the effective use of hyperparameter tuning.

There are five sections in this study. Following the introduction, the second section presents a review of the literature. In section three, the proposed section highlights the overview of the models used for detecting dementia is discussed. Section four details the experimental study and the results obtained. The conclusions and future work compose the last section.

2 Literature Review

Various studies performed use machine learning, ADL data, MRI data to provide innovative solutions for medical problems. An important element closely monitored in elderly patients is ADL. A vast array of experts collect and compile the relevant data which is further applied to the medical diagnosis of disorders. Wearable technologies are used to record and process ADL data. The machine learning models built-in [13] recorded accuracies of 90% showcasing the effective utility of algorithms. An association between Dietary Diversity score (DDS) and ADL utilized Logistic Regression (LR) models. A higher DDS value denoted to have a lesser chance of having an ADL disability [32]. The application of ADL data is further explored in the detection of dementia. Becker S. et al. [9] used correlations between depression, anxiety, sleep disturbances, and hallucinations or also known as DASH scores and Parkinson's Disease Questionnaire (PDQ) to find the risk of getting dementia. LR models tracked the trajectory of instrumental activities of daily living (iADL) in persons who have a higher risk of developing dementia. In [14], Cloutier et al. found quadratic functions can be used to indicate a decline just before the progression toward dementia. Lee et al. [21] found important associations between balance, gait, and ADL assessments in demented people. Factors such as step length, stride length, and the quality of gait are largely associated with the same. According to [30] in severely demented people, the cognitive function was significantly associated with the Personal Self-Maintenance Score (PSMS), which is used to assess ADL.

Machine learning is largely applied to MRI data in the medical community. MRI scans extract and use neuroimage-related information extensively in the diagnosis of brain-related diseases and disorders. In [19], Neuropsychiatric Inventory Questionnaire (NPI-Q) was used for finding the Mild Behavioral Impairment (MBI) status of the brain. An accuracy of 85% was recorded and distinguished between a normal and Alzheimer's patient. Other applications of MRI data include Significant Prostate Carcinoma (sPCA) classification. It was performed by [12] Castillo et al. using MRI data and different machine learning algorithms. Experts in the field apply relevant machine learning techniques to MRI data to detect and identify the progression of brain disorders in patients such as dementia. Sophisticated machine learning algorithms such ANNs, SVMs, and Adaptive neuro-fuzzy inference systems (ANFIS) were used [11] to differentiate VD and AD. Resting-state fMRI and Diffusion Tensor Imaging (DTI), some of the features from MRI were given as inputs to the models. ANFIS was able to achieve an accuracy of about 84%. Diagnosis evidence from a 3-year clinical follow-up was used to compare the model predictions.

Another approach to predicting the severity of dementia was witnessed in [8]. The authors used a Bag-of-Features (BOF) approach to extract features from MRI scans to classify the stages into demented, mildly cognitive impaired, and normal control. The machine learning model used multiclass SVM to give an accuracy of about 93%. Lian et al. designed a Disease Attention Map (DAM) and an attention-guided Hybrid Network (HybNet) consisting of sMRIs for AD classification which gave an accuracy of 82.7% with a sensitivity of 57.9% [22].

Identifying the most suitable form of the machine learning algorithm is a challenging task. A combination of models or algorithms showcases unique and distinct results. Ensemble models are a preferred choice as multiple models used in combination, strengthen the overall functionality and boost the overall effectiveness while being applied vastly in multiple domains. In [34], a machine vision system was designed to detect the surface defects on glass substrates of thin-film transistor liquid crystal displays. Ensemble classifiers such as Random Forest (RF) and Adaptive boosting models were used to handle the manufacturing objectives. AdaBoost exhibited better performance in comparison to the other models while detecting glass substrates. Another unique and effective application is seen in [16] where a heterogeneous ensemble-learning model was constructed for landslide susceptibility mapping using a spatial database. When compared to the individual classifiers, the blended ensemble method gives an overall accuracy of 80.7%. Medical science relies on intricate machine learning models or algorithms to assist in diagnosis. Ensemble modeling has vastly been applied and implemented across various sub-domains. Ensemble modeling is also used to classify medical data as observed in [7] using an R-Ensembler technique yielding superior results in comparison to NB and DT classifiers. Similarly, in [18] a Disease Prediction Model (DPM) was proposed. Outlier detection was performed using isolation forest and oversampling was done using SMOTE-Tomek. Hypertension risk was predicted using ensemble modeling, showcasing significant results.

Other medical applications such as multiple sclerosis prediction also used ensemble modeling. Simple SVM and LR models were compared with ensemble models like XGBoost and LightGBM. The models were built using the CLIMB dataset and were validated on the EPIC dataset; ensemble models sustained their performance on both models with relatively high-performance metrics [33]. Lastly, ensemble models showcased superlative results in the case of [20]. The authors developed a stacked ensemble classifier consisting of independently strong models with a meta learner. It was used for giving treatment of Cranial RadioTherapy for Acute Lymphoblastic Leukemia. A highly reasonable performance was obtained with an AUC value of 0.875.

Healthcare is a vast field of research. Ensemble machine learning models offer benefits and are a preferred medium of choice in the detection of dementia. The authors of [27] analyzed different ensemble models and concluded that the stacked model of GB and ANN yielded about 89% weighted average accuracy. Machine learning algorithms are applied to standard neuroimage datasets to detect dementia. In [25] the cross-sectional data of the OASIS dataset was analyzed using different machine learning techniques including SVM, GBM, and RF. Naidu et al. figured that the latter yielded higher performance metrics with minimum error. Another example of the utility of the OASIS dataset and ensemble models is seen in [15]. The authors compared the ensemble models with NB and ANN models. XGB was proposed as a fair choice to detect dementia in patients as it derived significant results with an accuracy of 97.8%.

Machine learning algorithms have been used extensively in the field of neuroscience. Detecting and predicting the onset of dementia and its stages is a critical and intricate task. Specialized machines are used in the medical field that capture scans to detect variations in body functions, and notice irregular patterns for analysis and diagnosis. It is a consistent effort requiring the patient to cooperate and participate in the tests developed by the specialists. Elderly care is essential as they form a part of the vulnerable population and require specialized treatments. They are unable to perform and remain at their best at all times due to a decline in aging factors. Thus, there is a system needed to identify the disorder through daily routine activities and further help a medical aid only when needed [28]. Using novel approaches helps develop solutions innovatively.

3   Proposed Multi-phase Detection of Dementia

In order to detect dementia at different stages, a multi-layered framework is proposed. The structure is designed to detect the presence or absence of dementia in an individual and the severity of the disorder is eventually predicted in the patient using additional neurological information. Figure 2 depicts an architecture diagram to showcase the overall process involved using 2 phases,

- In Phase 1, the application of ADL data predicts the presence or absence of dementia
- In Phase 2, when the patient is identified with symptoms of the disorder, an MRI scan is further used to predict the severity of dementia.

The primary focus of the study is to utilize routine activity information to specify whether an individual is dementia-prone or not at an initial stage itself. MRI scans are used as an additional medium to further expand on the disorder's impact on the individual.

A Center of Advanced Studies in Adaptive Systems (CASAS) dataset is used for dementia prediction in Phase 1. The cognitive assessment activity dataset comprises 10 input numeric variables and 1 target variable which is categorical. The target variable 'Dementia' oscillates between two values largely 0 and 1 determining the final state, i.e. if the patient is subject to the risk of dementia or not.

MRI scans are collated, categorized, and carefully organized by experts in the Open Access Series of Imaging Studies (OASIS) project which is used in Phase 2 of the study [23] The cross-sectional dataset or OASIS-1 comprises data related to 416 patients and OASIS-2 of 150 patients. A combination of features is used in the dementia severity detection process. Clinical Dementia Rating (CDR) is chosen as the target variable to identify the severity of the disorder predicted from Phase 1.

At both stages, relevant features are extracted to streamline the detection process. The datasets in the study undergo basic pre-processing techniques. Data has been scaled to ensure standardization across features. Random oversampling is done to maintain an equal number of samples in classes. In addition, MinMax scaling is performed to normalize the data. In the case of MRI data used, the null values are dropped and data encoding is performed for the 'gender' feature. Ensemble learning has been applied to the pre-processed datasets to yield suitable results.

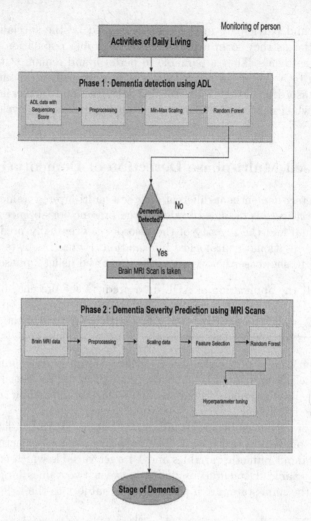

Fig. 2. Architecture diagram - multi-stage model for dementia detection and severity prediction

3.1   Phase 1 - Dementia Detection Using ADL Data

In Phase 1 of the study, a set of participants are monitored for a prolonged period of time and the routine activities are recorded using sensors. A set of tasks are given such as cooking, answering the telephone, etc. to be performed for a fixed time period. A score is devised based on the sequencing and execution of the tasks. Machine learning is applied to the datasets devised, assisting in the detection of the presence of dementia. The model is built using the scores for the tasks along with the sequencing score. Later when real-time ADL data is given to predict the score, the output determines whether the participant is subject to the risk of dementia or not. If the model predicts the onset of the illness, the caretaker is alerted to take an MRI scan and provide respective medications.

3.2   Phase 2 - Dementia Severity Prediction Using MRI Scans

In the study, after the detection of patients who are dementia prone, the individual's MRI scans are used to determine the severity of the illness. The dataset is curated by extracting common features from different scans of the brain. The MRI combinational dataset uses regular features such as gender, age, and MMSE scores. Unique features related to intracranial volume and whole-brain volume have been included as well. Further, a combination of relevant features has been carefully extracted and utilized to predict the stages of dementia. The severity is predicted accordingly using the chosen machine learning approach that provides computation benefits to design assistive devices.

3.3   Application of Random Forest Classifier in Phase 1 and 2 of Dementia Detection

The task of classification in machine learning is a decision-making process. Random Forest (RF) is an example of supervised classification algorithms. A combination of many decision trees selects features in the process of splitting. The decision trees produced showcase independent results and low correlations leading to a streamlined form of random guessing. RF demonstrates the concept of bagging or bootstrap aggregation in order to prevent excessive uncorrelation. The variance of the decision tree is reduced. Replacements are preferred for a sampling of data ensuring an overall diversity. Predictions generated from ensemble forms of learning are more accurate. The underlying concept is finding root nodes and splitting features randomly. When there are enough trees in the forest, overfitting is reduced [26]. The application of the RF classifier in the proposed framework is showcased,

Phase 1 - Use of RF for dementia detection

- Random samples of ADL data are picked from ADL dataset
- Each sample of ADL data generates a relevant decision tree, formulating a predicted result from each
- Predicted results undergo a voting process to classify the patient as dementia-prone or not
- Final prediction is estimated by the majority of votes supplied toward factors or features indicating the presence of dementia.

Phase 2 - Use of RF for dementia severity prediction

- Random samples of cross sectional and longitudinal MRI data are picked from OASIS's combinational preprocessed dataset
- Each sample of specialised combinational MRI data generates a relevant decision tree, formulating a predicted result from each

- Predicted results undergo a voting process to classify the severity of dementia in the dementia-prone patient identified in Phase 1
- Final prediction is estimated by the majority of votes applied toward factors or features indicating the severity of dementia of the patient from Phase 1.

4   Experimental Study

The following section comprises the different phases of the study performed to detect the presence of dementia using ADL data and predict the severity of the disorder using MRI scans as an additional source of information in the proposed multi-phase approach as follows,

4.1.   Analysis on Phase 1 Using ADL Data

Dataset: The ADL dataset is a specialized dataset developed by the researchers of Washington State University [3], which is a detailed study focusing on understanding how memory abilities share a relationship with simple activities such as cooking or the use of a telephone. The patients are exposed to a smart environment performing regular tasks in an orderly fashion. Essential details have been monitored and documented to generate the dataset. The values are represented numerically ranging between 0 and 5. In the paper, the values are utilized to analyze the risk of dementia in patients based on regular activities performed, the order of tasks helps understand the flow of activity. Any significant deviation helps indicate if the person's memory is in a progressive state of dementia or not.

The dataset comprises 8 scores related to indoor tasks, ratings for outdoor tasks, and sequencing scores for activities. Scores 1–8 are representative of the daily tasks designed for the patients to perform in the given environment. Simple day-to-day activities like sweeping the kitchen, getting a set of medications, and answering the phone call are scored based on the quality of work and the order of instructions carried out. Based on the memory, state of mind, or age, a patient can tend to be forgetful which is an indication of a unique scenario. The Sequencing Score field highlights the variation in the flow of activity of patients. If the activities are performed in the predetermined flow, the risk of dementia is mild, or even null.

Preprocessing: In order to organize the chosen data, preprocessing is performed. Records consisting of null values have been discarded. Scaling of features is performed to normalize all the values to a stipulated range. The non-null records are compiled to frame the final preprocessed dataset. The final preprocessed range of values is between 0 and which is done to ensure standardization across all values for ease of processing while applying the machine learning algorithms.

The first few records of the CASAS dataset that have been subjected to preprocessing are presented in Table 2.

Table 2. Phase 1 - Sample preprocessed ADL data

No.	S1	S2	S3	S4	S5	S6	S7	S8	Rating for outdoor tasks	Sequencing Score activities	Dementia risk
0	0.00	0.00	0.00	0.00	0.0	0.0	0.25	1.0000	0.30769	0.25571	1
1	0.25	0.00	0.00	0.00	0.0	0.0	0.25	0.3333	0.57692	0.21425	0
2	0.00	0.75	0.00	0.00	0.0	0.0	0.25	0.3333	0.26923	0.25571	0
3	0.00	0.00	0.00	0.25	0.0	0.0	1.0	0.3333	0.26923	0.25571	0
4	0.00	0.00	0.25	0.00	0.0	0.0	0.25	0.0000	0.19230	0.35714	0

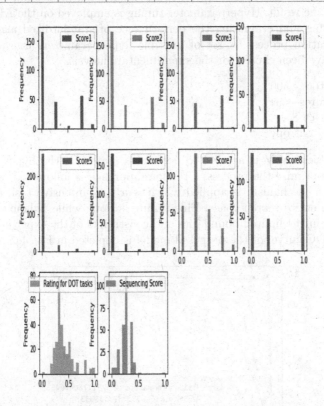

Fig. 3. Phase 1 - Histogram plot highlighting the different features incorporated

Feature Selection: Multiple features are used in the process of detection of dimension and the relevant distribution is represented in Fig. 3.

Evaluation and Analysis: In compliance with the proposed methodology, machine learning algorithms have been applied to derive the required results in both phases. Detection of dementia in Phase 1 is implemented by understanding the ADL dataset used and subjecting it to necessary preprocessing. The algorithms are applied to the preprocessed ADL data and the CASAS dataset provides the following critical information,

0 - represents the state where the patient is not subject to the risk of dementia or is non-demented

1 - represents the state where the patient is subject to the risk of dementia or has dementia.

The captured information is useful in identifying the different stages of dementia in Phase 2. Identification of whether the patient is subject to the risk of dementia or not is performed using an ensemble form of learning. Stratified sampling is applied using the StratifiedKFold mechanism with five folds. The process is useful to understand how the chosen input features combine effectively to generate the result. Hyperparameter tuning is employed on the ADL dataset while applying and evaluating the performance of standardized algorithms in the classification process. A set of multiple hyperparameters is available, the following have been chosen for the experimental analysis,

- n_estimators - 500
- max_features -sqrt
- max_depth - 8
- criterion - entropy

RandomisedSearchCV is applied to perform a search on the hyperparameters in order to optimize the process, a random approach is followed. Traditionally, a grid search mechanism is applied which is a time-intensive task as it scans through the entire search space. Fine-tuning is done while balancing multiple aspects within the limited boundaries and constraints of the experimental analysis. The ROC curve for the proposed model is depicted in Fig. 4.

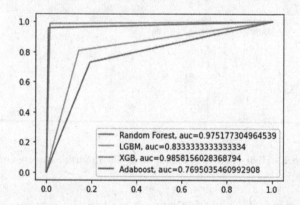

Fig. 4. ROC curve for the models predicting the risk of dementia using ADL data

RF classifier outperforms and satisfies the needs and requirements of the objective of Phase 1 of the experiment. Patients are subject to tests, health results are explored, and the accuracy needs to be as sharp as possible. Different ensemble models such as RF, XGBoost, LightGBM, and Adaboost are used. The results are depicted in Table 3 indicating the performance of RF is the most suited.

Table 3. Phase 1 - Evaluating results of ensemble models using ADL data

Model name	Accuracy	Precision	Recall	Std dev
Random Forest	95.74%	100%	91.48%	0.064
XGBoost	98.22%	97.88%	98.58%	0.071
LightGBM	78.72%	81.88%	73.75%	0.039
AdaBoost	79.07%	81.06%	75.88%	0.047

Hyperparameter tuning offers the benefit of building the model effectively in order to best extract the results while ensuring maximum accuracy possible with minimum error rate. In Fig. 5, the model employing Adaboost performs the least satisfactorily. The RF ensemble model for classification is able to maintain accuracy in the higher range, recording the highest overall performance in this phase of evaluation.

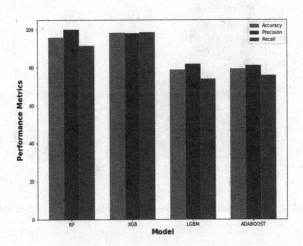

Fig. 5. Phase 1 - Overall comparison of ensemble model performances for dementia prediction

4.2   Analysis of Phase 2 Using MRI Data

Dataset: In order to proceed with the experiment, an additional set of dataset is required. Open source neuroimages are available for research and OASIS is a leading provider who contributes to the field extensively. Experts have recorded details of patients at different stages of dementia and shared the information on a public platform available for experimental analysis. In Phase 2, a unique dataset is curated as a combination of both cross-sectional and longitudinal neuroimagery data. According to standard medical practices, the progression of AD is measured by a factor known as Clinical Dementia Rating (CDR), which is chosen as the target variable. From the multitude of features offered, eight common features are extracted and preprocessed. The target variable chosen is variable in nature. Phase 2 detects the severity of a patient who has dementia using biomedical information. The process includes the application of machine learning algorithms on MRI based data to obtain the necessary results.

Preprocessing: The combinational data undergoes relevant preprocessing steps. Feature extraction is done to retain normalised and standard records for machine learning processing. Pre-processing is performed to filter out common features. Null records are discarded. Features are scaled to ensure uniformity. Figure 6 showcases a sample of the pre-processed MRI data. Relevant algorithms are applied to predict the severity of dementia with hyperparameter tuning. Results include the evaluation of the various models generated for the process of dementia severity prediction.

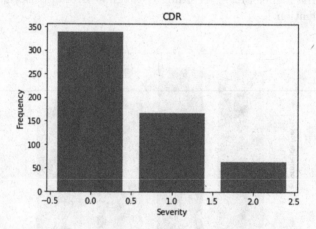

Fig. 6. Phase 2 - Target variable frequency depiction in combined MRI dataset

Feature Selection: Feature selection technique is applied to choose the common features required for further processing. The histogram in Fig. 7, showcases the eight factors taken into account after preprocessing for classification task of severity prediction. In order to further streamline the process of computation, features are selected strategically. Visualisation of the various features are mapped onto a histogram to showcase how the relationship varies with frequency.

Ideally, an experiment is performed to function at its best or detect an anomaly or outlier. A series of features are included in the process. There are methods and techniques to detect irrelevant features. The process of choosing the most appropriate features that assist in improving a model's performance is known as feature selection. In this study, feature selection is performed using SelectKBest technique. Statistical analysis is performed in the background to understand the relationship of the various features or variables with the target variable. The best features chosen are on the basis of the analysis.

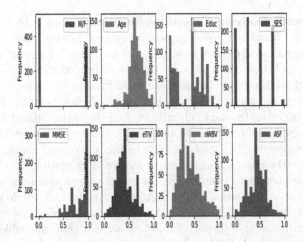

Fig. 7. Combination of different features of OASIS dataset

In SelectKBest technique it focuses on score functions in order to reduce the irrelevant features. This impacts the overall performance of the models. Filtering of the best features on the basis of highest 'k' scores, representing the number of top features to be selected, is performed in Phase 2 of evaluation. Five most important features namely - gender, age, SES, MMSE and nWBV are chosen. An overview of the features and their dependency is seen in Fig. 8. Basic features include gender, age and education indicated by years. Specialised features used include State Examination Score (SES), Estimated total intracranial volume (eTIV), Normalised whole-brain volume (nWBV), Atlas Scaling Factor (ASF) and CDR. CDR ranges between 0 and 2 with the following indications where 0 represents low level risk, 1 indicates mid-level risk and 2 denotes severe risk.

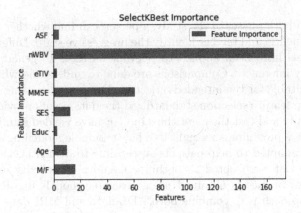

Fig. 8. Application of SelectKBest for feature selection on combined dataset

Evaluation and Analysis: Detection of dementia using both cross-sectional and longitudinal views is a novel approach explored in this phase of the experiment. The preprocessed dataset is a combination of both forms of views of neurological data and provides additional understanding of the stages of dementia which are used for the final classification task. Similar to Phase 1, the methodology is adapted for the experimental analysis of classification. Evaluation is performed considering the same algorithms and the standardized metrics.

Hyperparameters of both simple and complex ensemble models are fine-tuned to analyse and generate the final results which is performed using Randomized-SearchCV since the data is continuous in nature. The method utilizes a smaller set of hyperparameters to tune rather than scanning through the entire hyper-parameter space. It uses cross-validation to perform the task where the classifier, parameter distribution and number of folds are considered. A score and parameter are generated as the output. Table 4 showcases that ensemble models like Random Forest and LightGBM continue to yield better results in comparison to other models with the former outperforming, differing in precision rates. Figure 9 is a diagrammatic representation of the same.

Table 4. Phase 2 - Evaluating results of models using MRI data for dementia severity prediction

Model name	Accuracy	Precision	Recall	Std dev
Random Forest	90.29%	88.88%	89.23%	0.021%
K Neighbors	80.94%	78.81%	80.94%	0.049%
SVM	74.91%	72.63%	74.91%	0.051%
Logistic Regression	74.43%	70.43%	74.43%	0.036%
LightGBM	89.70%	89.32%	89.70%	0.023%

Thus, from both phases of the study, a person can find whether they are susceptible to dementia and further identify the progressive state. Different machine learning models have been applied to relevant datasets to generate results to draw necessary inferences. Comparisons are done to understand which approach is the most suitable for the intended objective. Data preprocessing consisting of feature scaling, feature selection standardized the data points providing a strong baseline to apply and test the algorithms on. RF have yielded significant results in both the cases providing an insight into how ensemble modelling and its application can be adapted to help patients susceptible to dementia. Deep Learning methods have not been opted to optimize memory complexity and since the equivalent performance is obtained using ensemble models like RF. The novelty of this approach is to combine both ADL data and MRI data to identify a demented person.

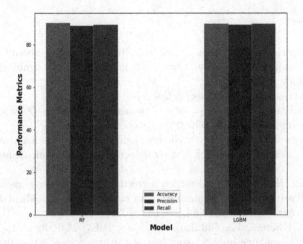

Fig. 9. Phase 2 - Overall comparison of ensemble model performances for dementia severity prediction

5   Conclusion

Aging is a natural process. Elderly individuals succumb to diseases that may require geriatric attention. Dementia impacts an individual's personality in severe conditions. There is a steady rise of cases and the progress in the medical field applies innovative methods to detect symptoms. As dementia has no cure and medicines are developed to help patients in preventing the progression of the illness. Diagnosis is a challenging process as symptoms can overlap which complicates the process further. Standardized assessment tests and tools are used widely to detect the disorder among the elderly. ADL and MRI data have been used individually to predict the occurrence of dementia in existing works which can be time, cost and resource intensive. In this paper, a novel attempt is made to detect dementia without MRI data and use it only to identify the severity of the illness. Daily routine activitiy data is used to identify consistent deviations and utilised effectively in identification of dementia-prone individuals in Phase 1. The proposed framework further used a distinct combination of neuroimage data to provide unique features in the process of dementia severity prediction. Machine learning offered simple and effective results with the adaptation of hyperparameter tuning and stratified folds. In both phases, ensemble models of learning yielded reliable results. The classification models generated were able to maintain an accuracy rate above 90% in both phases. There is scope and development of usage of more relevant forms of ADL or MRI data. Deep learning techniques can be applied to the study of dementia detection and severity prediction. In addition, a variety of physio parameters can also be applied to ensure more effectiveness in understanding the progress of the disorder, thus providing added benefits to geriatric care professionals.

References

1. About dementia. https://www.nhs.uk/conditions/dementia/about/
2. Cognitive impairment. https://www.cdc.gov/aging/pdf/cognitive_impairment/cogimp_poilicy_final.pdf
3. Datasets. https://casas.wsu.edu/datasets/
4. The global impact of dementia. https://www.alzint.org/u/WorldAlzheimerReport2015.pdf
5. What is dementia? https://www.cdc.gov/aging/dementia/index.html
6. What is dementia? https://www.dementia.org.au/about-dementia/what-is-dementia
7. Bania, R.K., Halder, A.: R-ensembler: a greedy rough set based ensemble attribute selection algorithm with knn imputation for classification of medical data. Comput. Methods Programs Biomed. **184**, 105122 (2020)
8. Bansal, D., Khanna, K., Chhikara, R., Dua, R.K., Malhotra, R.: Classification of magnetic resonance images using bag of features for detecting dementia. Proc. Comput. Sci. **167**, 131–137 (2020)
9. Becker, S., et al.: Association of cognitive activities of daily living (ADL) function and nonmotor burden in nondemented Parkinson's disease patients. Neuropsychology **34**(4), 447 (2020)
10. Breijyeh, Z., Karaman, R.: Comprehensive review on Alzheimer's disease: Causes and treatment. Molecules **25**(24), 5789 (2020)
11. Castellazzi, G., et al.: A machine learning approach for the differential diagnosis of Alzheimer and vascular dementia fed by MRI selected features. Front. Neuroinform. **14**, 25 (2020)
12. Castillo T, J.M., Arif, M., Niessen, W.J., Schoots, I.G., Veenland, J.F., et al.: Automated classification of significant prostate cancer on MRI: a systematic review on the performance of machine learning applications. Cancers **12**(6), 1606 (2020)
13. Chen, P.W., Baune, N.A., Zwir, I., Wang, J., Swamidass, V., Wong, A.W.: Measuring activities of daily living in stroke patients with motion machine learning algorithms: a pilot study. Int. J. Environ. Res. Public Health **18**(4) (2021). https://doi.org/10.3390/ijerph18041634, https://www.mdpi.com/1660-4601/18/4/1634
14. Cloutier, S., Chertkow, H., Kergoat, M.J., Gélinas, I., Gauthier, S., Belleville, S.: Trajectories of decline on instrumental activities of daily living prior to dementia in persons with mild cognitive impairment. Int. J. Geriatr. Psychiatry **36**(2), 314–323 (2021)
15. E, S.S.V., Shahina, A.N.K.: Dementia prediction on oasis dataset using supervised and ensemble learning techniques. Int. J. Eng. Adv. Technol. **10**(1), 244–254 (2020)
16. Fang, Z., Wang, Y., Peng, L., Hong, H.: A comparative study of heterogeneous ensemble-learning techniques for landslide susceptibility mapping. Int. J. Geogr. Inf. Sci. **35**, 1–27 (2020)
17. Ferrari, E., Cravello, L., Bonacina, M., Salmoiraghi, F., Magri, F.: Chapter 3.8 - stress and dementia. In: Steckler, T., Kalin, N., Reul, J. (eds.) Handbook of Stress and the Brain, Techniques in the Behavioral and Neural Sciences, vol. 15, pp. 357–370. Elsevier (2005). https://doi.org/10.1016/S0921-0709(05)80064-1, https://www.sciencedirect.com/science/article/pii/S0921070905800641
18. Fitriyani, N.L., Syafrudin, M., Alfian, G., Rhee, J.: Development of disease prediction model based on ensemble learning approach for diabetes and hypertension. IEEE Access **7**, 144777–144789 (2019)

19. Gill, S., et al.: Using machine learning to predict dementia from neuropsychiatric symptom and neuroimaging data. J. Alzheimer's Dis. (Preprint) **75**(1), 277–288 (2020)

20. Kashef, A., Khatibi, T., Mehrvar, A.: Prediction of cranial radiotherapy treatment in pediatric acute lymphoblastic leukemia patients using machine learning: a case study at MAHAK hospital. Asian Pac. J. Cancer Prev. **21**(11), 3211–3219 (2020)

21. Lee, N.G., Kang, T.W., Park, H.J.: Relationship between balance, gait, and activities of daily living in older adults with dementia. Geriatr. Orthop. Surg. Rehabil. **11**, 2151459320929578 (2020)

22. Lian, C., Liu, M., Pan, Y., Shen, D.: Attention-guided hybrid network for dementia diagnosis with structural MR images. IEEE Trans. Cyber. **52**(4), 1992–2003 (2020)

23. Marcus, D.S., Fotenos, A.F., Csernansky, J.G., Morris, J.C., Buckner, R.L.: Open access series of imaging studies: longitudinal MRI data in nondemented and demented older adults. J. Cogn. Neurosci. **22**(12), 2677–2684 (2010)

24. Murray, A.M., Seliger, S., Stendahl, J.C.: Chapter 21 - neurologic complications of chronic kidney disease. In: Kimmel, P.L., Rosenberg, M.E. (eds.) Chronic Renal Disease, pp. 249–265. Academic Press, San Diego (2015). https://doi.org/10.1016/B978-0-12-411602-3.00021-4, http://www.sciencedirect.com/science/article/pii/B9780124116023000214

25. Naidu, C., Kumar, D., Maheswari, N., Sivagami, M., Li, G.: Prediction of Alzheimer's disease using oasis dataset. Int. J. Recent Technol. Eng. **7**(6S3), 36–39 (2019)

26. Pedregosa, F., et al.: Scikit-learn: machine learning in Python. J. Mach. Learn. Res. **12**, 2825–2830 (2011)

27. Rawat, R.M., Akram, M., Pradeep, S.S., et al.: Dementia detection using machine learning by stacking models. In: 2020 5th International Conference on Communication and Electronics Systems (ICCES), pp. 849–854. IEEE (2020)

28. Ryu, S.E., Shin, D.H., Chung, K.: Prediction model of dementia risk based on XGBoost using derived variable extraction and hyper parameter optimization. IEEE Access **8**, 177708–177720 (2020)

29. Huber, S.J., Shuttleworth, E.C., Paulson, G.W.: Cortical vs subcortical dementia: neuropsychological similarities reply. Arch. Neurol. **44**(2), 131–141 (1987)

30. Tanaka, H., Nagata, Y., Ishimaru, D., Ogawa, Y., Fukuhara, K., Nishikawa, T.: Clinical factors associated with activities of daily living and their decline in patients with severe dementia. Psychogeriatrics **20**(3), 327–336 (2020)

31. Villa, C.: Biomarkers for Alzheimer's disease: where do we stand and where are we going? J. Personal. Med. **10**(4), 238 (2020)

32. Zhang, J., et al.: Beneficial effect of dietary diversity on the risk of disability in activities of daily living in adults: a prospective cohort study. Nutrients **12**(11) (2020). https://doi.org/10.3390/nu12113263, http://www.mdpi.com/2072-6643/12/11/3263'

33. Zhao, Y., et al.: Ensemble learning predicts multiple sclerosis disease course in the summit study. NPJ Digit. Med. **3**(1), 1–8 (2020)

34. Zuvela, P., Lovric, M., Yousefian-Jazi, A., Liu, J.J.: Ensemble learning approaches to data imbalance and competing objectives in design of an industrial machine vision system. Industr. Eng. Chem. Res. **59**(10), 4636–4645 (2020)

A Distributed Ensemble Machine Learning Technique for Emotion Classification from Vocal Cues

Bineetha Vijayan[✉], Gayathri Soman, M. V. Vivek, and M. V. Judy

Department of Computer Applications, Cochin University of Science and Technology, Kochi, Kerala 682022, India

{bineethavijayan1995,gayathrisoman0250,vivekmv, judynair}@cusat.ac.in

Abstract. Human-computer interaction and the creation of humanoid robots both depend heavily on emotions. By integrating the concept of emotion understanding, intelligent software systems become more effective and intuitive in resembling human-human interactions. Typically, we combine factors like intonation (speech), facial expression (visual modality), and word content (text). All possible multimodal combinations must be taken into consideration to process emotions appropriately. Among multimodal approaches, the use of human audio samples for emotion processing is given more weight than the use of facial expressions. To accomplish accurate categorization, analyzing massive volumes of real-time data has become more necessary. Machine Learning (ML) models that operate in a distributed fashion are crucial, given the size and complexity of the problem under study. In this respect, we propose a distributed ensemble model for vocal cue-based emotion classification. Three base ML models that work in a distributed manner were used. According to the findings, the ensemble model proposed differentiates between the seven fundamental emotions with reasonable accuracy. The proposed distributed ensemble model performed better than existing ML models on TESS, SAVEE, and RAVDESS, achieving 86% accuracy on the unified dataset.

Keywords: Emotion classification · Vocal cues · Distributed ML algorithms

1 Introduction

Every aspect of a person's life is influenced by their emotional state, including their decision-making, ability to perceive, relationships with others, intellect level, and physiological and psychological well-being [1]. In everyday interactions, people express their feelings through vocal, hand, face, and body movements, based on which the researchers have classified emotions differently [2]. The primary basic emotions observed in the majority of research studies include happy, sad, anger, fear, disgust, and surprise are the very first reaction of a person in a situation. Secondary emotions like anxiety, depression, and hate elicit a mental image or a feeling after experiencing the primary emotion. From another perspective, the emotions are mapped into a two-dimensional Valence-Arousal

P. P. Roy et al. (Eds.): BDA 2022 India, LNCS 13773, pp. 136–145, 2022.
https://doi.org/10.1007/978-3-031-24094-2_9

[3]. Normally when text communication fails to convey emotional states, speech communication is typically more evident and successful [4]. The emotional information concealed in speech contains both linguistic (syntactic-semantic) and paralinguistic (tone, pitch, intensity, etc.) contents; hence it is a crucial component of human contact and communication [5]. Significant effort has been put into ML methods to realize emotion classification based on speech. To overcome the challenge of larger datasets, we might use a distributed environment optimized for a deep ML method.

The following are the significant contributions of the paper:

- A distributed ensemble model for emotion classification from vocal cues is proposed which performs better than other approaches for emotion classification from vocal cues when processing time is taken into account.
- A unified dataset was also built from the three datasets after certain pre-processing stages. This can be used for future research in which large-scale learning is required.
- The internal job execution of different ML algorithms using the Spark environment is explained with the help of a Directed Acyclic Graph (DAG).

In this work, we focused on training ML models in a distributed environment with the goal of reducing training time by ensuring that the model is trained on the correct set of parameters. This paper is organized by including the following. Section 2 presented the related research on the use of ML for emotion detection using speech samples. Section 3 details the proposed framework and the audio preprocessing procedure, spectral feature extraction and reduction mechanisms. Finally, Sect. 4, discusses the various methods used to develop the ML models and the outcomes attained when using distributed learning strategy to do discrete emotion classification. Later concluded the paper by giving insights into the future work possible in this domain.

2 Related Works

As far as psychological concepts are concerned, emotion is up there with the most elusive ones. In emotional computing, there are two sorts of modelling emotion: discrete emotion models [6] and dimensional emotion models [7, 8]. Ekman's six fundamental emotions [6] and Plutchik's emotional wheel model [9] are two extensively utilized emotion models. The underlying emotions for each utterance are classified using the speech emotion recognition (SER) system. SER systems often make use of a plethora of traditional classification algorithms. Gaussian mixture model (GMM) classifiers were used to categorize speech based on their expressions of the five primary emotions [10]. Hidden Markov Model (HMM) fares better on the logarithmic frequency power coefficient for text-independent emotion classification with a recognition rate of 89.2% [11]. Y. Pan et al. [12] employed a support vector machines (SVM) classifier to categorize three emotions, sad, joyful, and neutral, trained on two emotional databases. SER solution proposed [13] aided children diagnosed with autism spectrum disorder (ASD) to identify emotions. Later the researchers [14] improved the performance of the classification by introducing an ensemble learning approach from three ML classifiers. Seven emotions were classified using an SVM-based binary decision tree in [15]. A study

[16] trained the Naïve Bayes classifier on EmoDB to recognize emotions. The authors coupled MFCC and pitch to improve SER's performance. Recently published review article studied the application of ML and deep learning (DL) algorithms to classify emotions in speech corpus [17, 18]. Long lin [19] suggested a multi-distributed SER system dependet on the mel frequency spectrogram and parameter transfer. A novel type of data augmentation network, called an adversarial data augmentation network (ADAN), was proposed by researchers Lu Yi et al. [20]. Even though there are a large number of audio data emotion processing algorithms already available, this is the first study to classify emotional audio content using a distributed ensemble model that we are aware of.

3 Proposed Framework

The proposed system is a distributed ensemble ML models for emotion classification from vocal cues, the general structure is presented in Fig. 1.

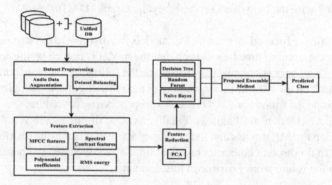

Fig. 1. General structure of the suggested ensemble model.

3.1 Dataset

The publically available speech databases created for emotion detection falls under three categories: spontaneous, simulated and elicited speech. In multimodal emotion detection scenarios, spontaneous speech [21] are well suited as the emotions are natural since the actors don't know that their motions are captured. Elicited speech [22] contains significantly mild emotions as they are induced by giving an artificial emotional situation with the actor's knowledge that they are being recorded. Above 60% SER system is trained using simulated databases. Hence in this experiment, the distributed ML models were trained on three simulated public databases; namely, toronto emotional speech set (TESS) [23], surrey audio-visual expressed emotion (SAVEE) [24] and ryerson audio-visual database of emotional speech and songs (RAVDESS) [25] were the samples are available in.wav format captured in time domain. The training time of ML models grows significantly when the data size is large [26]. In order to demonstrate the usage of distributed ML algorithms, a unified dataset was built by combining the aforementioned datasets and is named "Unified DB". The unified dataset has a size of 926 MB.

3.2 Preprocessing

The initial exploratory data analysis on the collected samples shows that the datasets are imbalanced. So, using the raw audio augmentation technique in MATLAB, the sample sizes of the classes which were minority resampled to match those of the majority classes. The following mapping was done for each emotion class, converting them into numerical labels in the range of 0 to 8: 01 = neutral, 02 = calm, 03 = happy, 04 = sad, 05 = angry, 06 = fearful, 07 = disgust, 08 = surprised. The "calm" class of the RAVDESS dataset is not used for this experiment. Table 1. Depicts the amount of samples in each emotion class of the datasets prior to and following dataset balancing.

Table 1. No. of samples in the datasets before and after balancing

Dataset before balancing	Emotions								Dataset after balancing	Emotions						
	01	02	03	04	05	06	07	08		01	03	04	05	06	07	08
RAVDESS	96	192	192	192	192	192	192	192	RAVDESS	192	192	192	192	192	192	192
SAVEE	120	0	60	60	60	60	60	60	SAVEE	60	60	60	60	60	60	60
TESS	400	0	400	400	400	400	400	400	TESS	400	400	400	400	400	400	400
Unified DB	616	192	652	652	652	652	652	652	Unified DB	652	652	652	652	652	652	652

3.3 Feature Extraction and Reduction

Features of human speech are the most prominent characteristics of any SER system. In spite of the proliferation of SER frameworks, there remains no agreed-upon set of features that may be used for systematic classification. The extraction of global or local features, or both, may be performed based on our requirements. In this work, the following spectral features were extracted to build the custom feature set as they generally have higher classification accuracy than others.

Mel Frequency Cepstrum Coefficients (MFCC) Features: The MFCC feature introduced in [27] is considered essential to distinguish various emotions from human speech. Here, the mean and standard deviation (SD) of the MFCC feature are used [14], because only focusing at the mean value tends to make the results less accurate. We measured the mean and standard deviations of the first 26 MFCCs for each audio frame.

Spectral Contrast Features: These features [28] constitute the relative spectral distribution including the spectral centroid, spectral bandwidth, spectral contrast, and many others. Here seven unique values of the same were extracted.

Polynomial Coefficients: Determines the coefficients of the nth-order polynomial that best fits the columns of a spectrogram by using the coefficients. Here two polynomial coefficients for a polynomial of order one were extracted per frame.

Root Mean Squared Energy: The root-mean-square (RMS) energy is extracted from each of the audio frame from the given audio samples.

Thus the resultant custom feature set consists of 62 low-level descriptors retrieved from the speech data by using a python toolkit named Librosa. Researchers have used Liborsa for various audio classification projects since it is a dependable technique for extracting audio features. Each feature is separately recorded for each individual dataset and for the unified DB using a sampling frequency of 16 kHz and saved as comma separated files (CSV). In order to guarantee the accuracy of computation, the values are rescaled to a uniform range.

Feature Reduction: The feature extraction process results in producing a plethora of features that need to be reduced to overcome the curse of dimensionality. For feature reduction principal component analysis (PCA) [29] was employed. The feature extraction technique produced 62 customized feature sets, fed into PCA to decrease the feature vector. In order to account for 95% of the overall variation, a total of 27–35 principal components is required.

3.4 Distributed Machine Learning Algorithms

Multinode ML methods and systems that are intended to enhance performance, boost accuracy, and scale to bigger input data quantities are referred to as distributed machine learning. Due to the scalability and efficiency issues with ML methods, handling big amounts of data can be difficult. For instance, the technique will not scale properly due to memory constraints when the algorithm's computational complexity exceeds the main memory. Due to their capacity to distribute learning processes across a number of workstations, distributed ML algorithms [26] are essential to large-scale learning. We employed the Spark distributed ML framework to create a scalable ML data pipeline. Spark is regarded as a quick, user-friendly, and all-purpose engine for massive data processing. Large volumes of data are processed and analyzed using a distributed computing engine. The ML models we used were decision tree, random forest, and Naïve Bayes. The model's performance was enhanced even more by using SparkML pipeline.

Decision Tree Classifier. Popular approaches for classification and regression challenges in ML include decision trees and their ensembles. We used the distributed spark decision tree method. The decision tree approach computes histograms simultaneously for all nodes at each tree level to speed up processing. Deeper levels of the tree could require a lot of memory as a result, which could result in memory overflow errors. To address this issue, the distributed decision tree approach uses a training parameter that determines the maximum amount of memory at the workers (and twice as much at the master) to be dedicated to the histogram computation. Once the memory requirements for a level-wise computation exceed the predetermined threshold, the node training tasks at each succeeding level are divided into smaller tasks.

Random Forest Classifier. Random forests are groups of Decision trees that Spark MLlib can handle. Multiple trees can be learned simultaneously in a Random Forest since each tree is learned individually (in addition to the parallelization for single trees).

A number of parallel subtrees are trained, and the number is adjusted depending on memory restrictions after each iteration. Each tree is trained in a random forest using a distinct subsample of the data. Instead of directly copying data, we conserve memory by employing a TreePoint structure that keeps track of how many copies of each instance are present in each subsample.

Naive Bayes Classifier. A straightforward multiclass classification algorithm that assumes each pair of features as independent. It employs Bayes' theorem to training data in a single pass, estimating the conditional probability distribution of each feature given label, and then utilizing that knowledge to predict the label given an observation. Spark MLlib supports both Bernoulli naive Bayes and multinomial naive Bayes.

Distributed Average-Based Ensemble Model. The results from an ensemble technique can be more accurate than those from a single model. The ensemble model's initial stage is the creation of base learners. An essential step in the ensemble technique is choosing the component classifiers based on their effectiveness. Combining individual classifier predictions is the second crucial step in ensemble learning, which is done using various techniques. In the proposed method we used average-based ensemble technique. In this method, the average of the class probabilities were taken into consideration. Let $Cn = [Pn1, Pn2...Pnm]$, be the class probability vector of n classifiers. In the proposed method, the average of the probability value of each classifier for a particular class was taken. Let $Avm = [P1m + P2m +... + Pnm]/n$, be the average of the probability value of n classifiers for a class m. The predicted class label is the class that has the maximum average value. Let Ci be the predicted class label for a sample, then $Ci = max (Av1, Av2...Avm)$, where Avm is the average of the probability value of n classifiers for a class m.

4   Experimental Setup and Analysis

This study was conducted in the data science research lab of computer applications department, CUSAT,Kochi, Kerala, using a high-performance Hadoop cluster. The cluster's servers have a total of 768 GB of RAM and 144 core processors, split between 1 Name node and 2 Data node servers. The Cluster is compatible with the Hortonworks Data Platform 3.0 (HDP 3.0) and Apache Spark 3.0.0. The spark platform generates a Directed Acyclic Graph that tracks all operations performed by the spark engine.

4.1   Results

The emotion datasets are divided into a training set comprising 75% of the data and a testing set comprising 25% of the data before being used to train ML models. Each ML model is individually trained based on hyperparameter configurations created. Model selection, or using data to choose the optimal Model or parameters for a specific job (Tuning), is an important issue in machine learning. CrossValidator and TrainValidationSplit help MLlib pick the best models.

Hyperparamter Tuning Using Grid Search and Cross-Validation. Grid search was used on each ML method to get the optimal hyperparameters from a set for model training. The best parameters of each ML model when trained on a unified DB is given below. Here to choose the best parameter values 5- fold cross-validation is conducted for tuning the paramGrid. The results after tuning is as follows. Best Parameters of Decision Tree: {maxDepth: 30, maxBin: 20). Best Parameters of Random Forest: {(maxDepth: 10, maxBin: 20, NumTrees: 50}.Best Parameters of Naive Bayes:{ smoothing: 0.4}. These parameters can be used for future executions for getting better classification accuracy. The Fig. 2. Shows the improvement in accuracy after applying grid search on TESS and Unified DB dataset.

Fig.2. Comparison of different ML model performances on TESS and Unified DB datasets before and after performing grid search.

Performance Comparison of the Proposed Model on Different Evaluation Metrics. The Table 2. Shows the metrics obtained by the distributed models decision tree, Naive Bayes, random forest, and the proposed average based ensemble model after testing on the unified DB. Results shows the suggested ensemble classifier has the highest performance compared to the three existing classifiers. This occurs since the misclassified samples of ML classifiers get correctly classified when applied on the ensemble classifier. By doing so, it reduces the issue of overfitting, lowers the variance, and ultimately increases accuracy.

Table 2. Performance comparison of proposed model based on distinct evaluation metrics.

Algorithm	Accuracy	Precision	Recall	F1 Score
Proposed	86	85.6	85.6	85.4
Decision tree	76	76.2	75.9	75.9
Random forest	85	82.5	81.9	81.8
Naïve Bayes	65	68.2	64.7	65.2

The directed acyclic graphs (DAG) are used to illustrate the distributed execution of the different ML algorithms in use. In Apache Spark, a DAG is made up of a set of vertices and edges, where the vertices stand in for RDDs and the edges for operations to be performed on RDD. In complex jobs, the significance of the DAG visualization is most obvious. The DAG visualization gives users the ability to click into a stage and enlarge its information as the timeline view does.

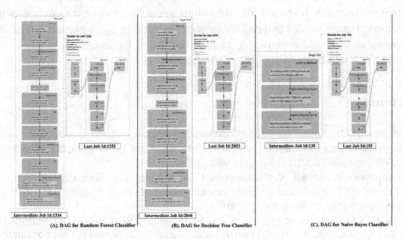

Fig.3. DAG illustrating the distributed execution of ML classifiers.

DAG generated for ML Classifier. Figure 3 Shows the DAG created for distributed ML algorithms fed with unified DB dataset. The compiler will load the dataset and distribute it across different nodes in Hadoop distributed file system. Using vector assembler the class label and its corresponding feature vectors are transformed to a key value pair. In the next stage, using a hash function, similar class labels and its feature vector will get grouped. Till this phase, it's similar for all ML algorithms. After that, the mapping function differs based on the architecture of ML models. For random forest classifier a total of 1352 jobs were created and mapped. Similarly decision tree task is distributed over 2053 jobs whereas for Naïve Bayes 155 jobs were created and distributed. The no. of jobs created will get distributed over 58 cores of the spark cluster in use. Finally, the results of the map functions are then fed into the reduce function, which summarizes the results based on the employed ML techniques. For example in case of Naive Bayes classifier, in the reduce function, the conditional probability is calculated and a conditional probability table is created for aggregating the results after distributed classification. By introducing distributed ML algorithms training of large scale data can be done with less computational complexity.

5   Conclusion

In this study, a distributed ensemble model is employed to categorize emotions based on vocal cues. The proposed method is tested on 3 datasets (RAVDESS, SAVEE, TESS) as well as on the unified DB created. The suggested network uses the SparkML pipeline technique for emotion classification. The proposed method can operate effectively on any larger dataset because of the utilization of SPARK distributed computing. The proposed distributed ensemble network excels the other individual classifiers because it calculates the final prediction by averaging the probability for each class, thus giving more importance to the class to which the sample actually belongs. With the aid of the DAG, the internal job execution of several ML algorithms in the Spark environment is explained in this work. Future studies that call for large-scale learning can make advantage of the uniform dataset to build even bigger dataset. The use of the model for real-time vocal cue processing will be the main goal of our future work.

References

1. Setyohadi, D.B., Kusrohmaniah, S., Gunawan, S.B., Pranowo, P., Prabuwono, A.S.: Galvanic skin response data classification for emotion detection. Int. J. Electr. Comput. Eng. **8**(5), 4004 (2018). https://doi.org/10.11591/ijece.v8i5.pp4004-4014
2. Goshvarpour, A., Abbasi, A., Goshvarpour, A.: An accurate emotion recognition system using ECG and GSR signals and matching pursuit method. Biomed. J. **40**(6), 355–368 (2017). https://doi.org/10.1016/j.bj.2017.11.001
3. Alarcão, S.M., Fonseca, M.J.: Emotions recognition using EEG signals: a survey. IEEE Trans. Affect. Comput. **10**(3), 374–393 (2019). https://doi.org/10.1109/TAFFC.2017.2714671
4. Emotion Recognition based on Heart Rate and Skin Conductance. In: Proceedings of the 2nd International Conference on Physiological Computing Systems, pp. 26–32 (2015). https://doi.org/10.5220/0005241100260032
5. Cowie, R., et al.: Emotion recognition in human-computer interaction. IEEE Signal Process. Mag. **18**(1), 32–80 (2001). https://doi.org/10.1109/79.911197
6. Ekman, P., et al.: Universals and cultural differences in the judgments of facial expressions of emotion. J. Pers. Soc. Psychol. **53**(4), 712–717 (1987). https://doi.org/10.1037/0022-3514.53.4.712
7. Mehrabian, A.: Basic Dimensions for a General Psychological Theory: Implications for Personality, Social, Environmental, and Developmental Studies, 1st ed., vol. 1. Oelgeschlager, Gunn & Hain, 1980 (1980)
8. Russell, J.A.: A circumplex model of affect. J. Pers. Soc. Psychol. **39**(6), 1161–1178 (1980). https://doi.org/10.1037/h0077714
9. Plutchik, R.: A general psychoevolutionary theory of emotion. In: Theories of Emotion, Elsevier, pp. 3–33 (1980)
10. Andrews, M.: Emotion recognition in spontaneous speech. In: Faculty Research Working Paper Series, vol. 28, pp. 457–473 (2014). https://bsc.cid.harvard.edu/files/bsc/files/285_and rews_this_is_pfm.pdf
11. Nwe, T.L., Foo, S.W., De Silva, L.C.: Detection of stress and emotion in speech using traditional and FFT based log energy features. In: Fourth International Conference on Information, Communications and Signal Processing, 2003 and the Fourth Pacific Rim Conference on Multimedia. Proceedings of the 2003 Joint, vol. 3, pp. 1619–1623 (*2003*). https://doi.org/10.1109/ICICS.2003.1292741
12. Pan, Y., Shen, P., Shen, L.: Speech emotion recognition using support vector machine. Int. J. Smart Home **6**(2), 101–108 (2012). https://doi.org/10.30534/ijeter/2020/43842020
13. Matin, R., Valles, D.: A speech emotion recognition solution-based on support vector machine for children with autism spectrum disorder to help identify human emotions. In: 2020 Intermountain Engineering, Technology and Computing (IETC), pp. 1–6 (2020). https://doi.org/10.1109/IETC47856.2020.9249147

14. Valles, D., Matin, R.: An Audio processing approach using ensemble learning for speech-emotion recognition for children with ASD. In: 2021 IEEE World AI IoT Congress (AIIoT), pp. 0055–0061 (2021). https://doi.org/10.1109/AIIoT52608.2021.9454174

15. Yuncu, E., Hacihabiboglu, H., Bozsahin, C.: Automatic speech emotion recognition using auditory models with binary decision tree and SVM. In: 2014 22nd International Conference on Pattern Recognition, pp. 773–778 (2014). https://doi.org/10.1109/ICPR.2014.143

16. Khan, A., Roy, U.K.: Emotion recognition using prosodie and spectral features of speech and Naïve Bayes classifier. In: 2017 International Conference on Wireless Communications, Signal Processing and Networking (WiSPNET), pp. 1017–1021 (2017). https://doi.org/10.1109/WiSPNET.2017.8299916

17. Prajapati, Y.J., Gandhi, P.P., Degadwala, S.: A review - ML and DL classifiers for emotion detection in audio and speech data. In: 2022 International Conference on Inventive Computation Technologies (ICICT), pp. 63–69 (2022). https://doi.org/10.1109/ICICT54344.2022.9850614

18. Wani, T.M., Gunawan, T.S., Qadri, S.A.A., Kartiwi, M., Ambikairajah, E.: A comprehensive review of speech emotion recognition systems. IEEE Access **9**, 47795–47814 (2021). https://doi.org/10.1109/ACCESS.2021.3068045

19. Lin, L., Tan, L.: Multi-distributed speech emotion recognition based on mel frequency cepstogram and parameter transfer. Chinese J. Electron. **31**(1), 155–167 (2022). https://doi.org/10.1049/cje.2020.00.080

20. Yi, L., Mak, M.-W.: Improving speech emotion recognition with adversarial data augmentation network. IEEE Trans. Neural Networks Learn. Syst. **33**(1), 172–184 (2022). https://doi.org/10.1109/TNNLS.2020.3027600

21. Cao, H., Verma, R., Nenkova, A.: Speaker-sensitive emotion recognition via ranking: Studies on acted and spontaneous speech. Comput. Speech Lang. **29**(1), 186–202 (2015). https://doi.org/10.1016/j.csl.2014.01.003

22. Basu, S., Chakraborty, J., Bag, A., Aftabuddin, M.: A review on emotion recognition using speech. In: 2017 International Conference on Inventive Communication and Computational Technologies (ICICCT), pp. 109–114 (2017). https://doi.org/10.1109/ICICCT.2017.7975169

23. Pichora-Fuller, M.K.. Dupuis, K.: Toronto emotional speech set (TESS). Borealis (2010).https://doi.org/10.5683/SP2/E8H2MF

24. Jackson, P.: Surrey audio-visual expressed emotion (SAVEE) database. http://kahlan.eps.surrey.ac.uk/savee/

25. Livingstone, S.R., Russo, F.A.: The ryerson audio-visual database of emotional speech and song (RAVDESS): a dynamic, multimodal set of facial and vocal expressions in North American English. PLoS ONE **13**(5), e0196391 (2018). https://doi.org/10.1371/journal.pone.0196391

26. Verbraeken, J., Wolting, M., Katzy, J., Kloppenburg, J., Verbelen, T., Rellermeyer, J.S.: A survey on distributed machine learning. ACM Comput. Surv. **53**(2), 1–33 (2021). https://doi.org/10.1145/3377454

27. Davis, S., Mermelstein, P.: Comparison of parametric representations for monosyllabic word recognition in continuously spoken sentences. IEEE Trans. Acoust. **28**(4), 357–366 (1980). https://doi.org/10.1109/TASSP.1980.1163420

28. Jiang, D.N., Lu, L., Zhang, H.J., Tao, J.H., Cai, L. H.: Music type classification by spectral contrast feature. In: Proceedings. IEEE International Conference on Multimedia and Expo, pp. 113–116. https://doi.org/10.1109/ICME.2002.1035731

29. Wold, S., Esbensen, K., Geladi, P.: Principal component analysis. Chemom. Intell. Lab. Syst. **2**(1–3), 37–52 (1987). https://doi.org/10.1016/0169-7439(87)80084-9

Graph Analytics

Drugomics: Knowledge Graph & AI to Construct Physicians' Brain Digital Twin to Prevent Drug Side-Effects and Patient Harm

Asoke K. Talukder[1,2]($\boxtimes$), Erwin Selg[3], Ryan Fernandez[4], Tony D. S. Raj[4], Abijeet V. Waghmare[4], and Roland E. Haas[5]

[1] SRIT India, 113/1B ITPL Road, Brookfield, Bangalore 560037, India
asoke.talukder@sritindia.com

[2] CSE, National Institute of Technology Karnataka, Surathkal, India

[3] SRH Fernhochschule GmbH, Kirchstraße 26, 88499 Riedlingen, Germany
erwin.selg@mobile-university.de

[4] Division of Medical Informatics, St. John's Research Institute, St. John's Medical College, Sarjapur Road, Bangalore 560034, India
{ryan,tonyraj,abijeet}@sjri.res.in

[5] International Institute of Information Technology Bangalore (IIIT-B), Bangalore, India
roland.haas@iiitb.ac.in

Abstract. Unintended toxic effects of a medication occur due to drug-drug interactions (DDI) and drug-disease interactions (DDSI). It is the fourth leading cause of death in the US. To overcome this crisis, we have constructed the Drugomics knowledge graphs comprising DDI and DDSI interactions mined from Drugs@FDA, FAERS (FDA Adverse Events Reporting System), PubMed, literature, DailyMed, drug ontology, and other biomedical data sources. We used Artificial Intelligence and Augmented Intelligence (AI&AI) to translocate this actionable DDI and DDSI knowledge into a network and stored it in a Neo4j property graph database in a cloud for anytime-anywhere access. For the first time, we present here an AI-driven Evidence-Based Clinical Decision Support (AIdEB-CDS) system that accepts human understandable plain text inputs and extracts knowledge from knowledge graphs to offer the right therapeutics for the right disease for the right person at the right time at any Point-of-Care. This functions like a physicians' brain digital twin to reduce clinical errors, reduce medication errors, and increase general health equity at a reduced cost. This will eliminate the patient harm caused by drug interactions,

Keywords: Patient harm · Drugomics · Drug-drug interaction · Drug-disease Interaction · Conflicting drug · Brain digital twins · DDI · DDSI

1 Introduction

A drug is transported by the body from the site of administration to the desired site of action to counter a disease or an illness. In this drug pathways, interactions take place

P. P. Roy et al. (Eds.): BDA 2022 India, LNCS 13773, pp. 149–158, 2022.
https://doi.org/10.1007/978-3-031-24094-2_10

with the comorbid diseases or other drugs already administered. Drug-drug interactions (DDI) occur when the pharmacologic effect of one drug is altered by the action of another drug already consumed by the patient in polypharmacy cases. Drug-disease interactions (DDSIs) in contrast are situations where the pharmacotherapy used to treat one disease conflicts with another multimorbid disease the patient is already suffering from. These interactions or conflicts cause unpredictable clinical effects and patient harm. Drug conflict or drug side-effect is the fourth leading cause of death in the US [1]. At the WHO Second Global Ministerial Patient Safety Summit in 2017, the "Medication Without Harm" program was launched. The aim of this program was to reduce the "avoidable medication-related harm" by 50%, from $42 billion to $21 billion globally by March 2022 [2]. In fact, the cost of medication-associated harm increased–in 2021 it exceeded $40 billion in the US alone [3].

To counter this avoidable patient harm, we have constructed a few knowledge graphs namely, Symptomatics, Diseasomics and Drugomics. Symptomatics [4] and Diseasomics [5] are discussed elsewhere. Drugomics presented here is to empower a doctor choose the right medication for the right disease and for the right patient, at the right time. Drugomics is constructed out of two major knowledge domains, namely, drug-drug interactions (DDI), and drug-disease interaction (DDSI). The drug-disease interaction can further be divided into drug-disease indications and drug-disease contraindications. This is the first time such computer-understandable holistic actionable drug knowledge that combines drug-drug interactions, drug-disease interactions, and drug-disease contraindications is democratized and translocated from evidence to a knowledge graph. This translocated holistic drug knowledge is the digital twin of physicians' brain. This in fact is a physicians' brain digital twin. This brain digital twin is accessible through REST/JSON-RPC API over the Internet for ingestion at any point-of-care.

This paper on physicians' brain digital twin to prevent drug related patient harm starts with an Introduction in Sect. 1. In Sect. 2, we discuss Drug-Drug Interactions (DDI) knowledge sources. In Sect. 3, we discuss Drug-Disease Interaction (DDSI) knowledge sources. In Sect. 4, we present the Drugomics Knowledge Graph constructed from DDI and DDSI data. In Sect. 5, we describe AI-driven Evidence-Based Clinical Decision Support (AIdEB-CDS) as a Use Case of Drugomics and medication without harm. We conclude the paper in Sect. 6.

2 Drug-Drug Interaction (DDI) Knowledge Sources

DDI information is limited and dispersed in scientific literature and labels published by pharmaceutical companies. For example, DrugBank [6] is a database containing comprehensive molecular information about drugs, but it is not publicly available for download. Some authors used text embedding on DrugBank data to discover the DDI [7]. Some Clinical Decision Support (CDS) systems used Medical Logic Module (MLM) using Arden syntax to represent and exchange medical information [8]. Limited adverse events are listed at the FDA site Drugs@FDA [9]. Stanford Network Analysis Project (SNAP) [10] offers interactions between FDA approved drugs. Tatonetti and team constructed an outstanding DDI database [11]. Xiong et al. published a dataset of drug-drug interactions [12]. Such systems work on localized knowledge and fail to provide holistic patient-level conflicts for both DDI and DDSI at the Point-of-Care.

As a clinical trial always focuses on the positive impact of the medications, they fail to obtain the entire range of possible DDIs [13]. These reactions are usually known only after the drug is released for clinical use. The FDA adverse event reporting system (FAERS) contains such post-marketing incidents in plain text [14]. We used all these open data for our DDI knowledge graph. All these drug-related data sources use different nomenclatures and terminologies to represent the information. Many of these information sources use proprietary standards. We normalized all these heterogenous terminologies and converted them into RxNORM [15] controlled vocabulary.

3 Drug-Disease Interaction (DDSI) Knowledge Sources

Drug-disease interactions (DDSI) are situations where the drug used to treat one disease causes worsening impact on another disease already present in the patient. In elderly population, 15%–16% of the patients had at least one drug-disease interaction. Managing the DDSI is essential to prevent serious patient harm or death [16].

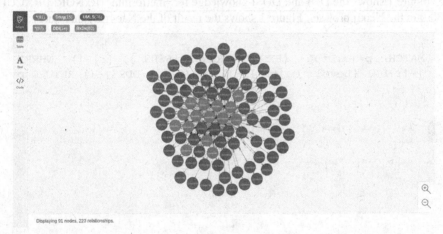

Displaying 91 nodes, 223 relationships.

Fig. 1. The drug-drug interaction and drug-disease interaction knowledge graph for 'Metformin' (RxNORM:6809) drug on Neo4j graph database browser.

Like the drug-drug interaction, drug disease interactions knowledge is not available in a central location. They are scattered and distributed across multiple sources in different formats. DailyMed [17] is the information resource provided by the US FDA that contains drug labels. There is an important work by Sharp on drug ontology that includes both drug indication and contraindication information [18]. Some drug interaction knowledge is also available in PubMed. We mined all these data sources. We normalized drug names and disease names into RxNORM and UMLS respectively to ensure seamless integration of DDI and DDSI knowledge.

4 Drugomics Knowledge Graph

Drugomics relates to holistic drug interaction knowledge. This includes the intended function of the drug (drug indication). The unintended toxic effects of the drug due to drug disease interaction (drug contraindication) and the drug-drug interactions. To achieve the systems biomedical view, we used antireductionistic approach. We normalized the knowledge and interconnected them through common RxNORM nodes. We wrote Neo4j Cypher commands to load them into the Neo4j graph database.

We took all these publicly available data cited above [9–12, 14, 16–18] step-by-step to construct the drugomics knowledge graph. Most of these data are unstructured and are available in plain text. We used NLP (Natural Language Processing) and other statistical algorithms to extract knowledge using semi-automated procedures. From this we identified total 6,684 drugs that participate in either DDI or DDSI interactions. We discovered total 433,043 interactions – out of which 319,514 are DDI relationships and 113,529 are DDSI relationships. We validated the knowledge graph by randomly selecting nodes through manual verification by expert doctors and pharmacists. We translocated this knowledge into a Neo4j property graph database [19]. We deployed this drug knowledge graph in a cloud as physicians' brain digital twin and made it accessible through Web REST/JSON-RPC API.

Figure 1 shows the DDI and DDSI knowledge for 'metformin' (RxNORM/RXCUI 6809) in the Neo4j browser. Figure 1 shows the result of the Neo4j Cypher query:

```
MATCH  p=(d:Drug  {RXNORM:'RXNORM:6809'})-[r]-()  WHERE
(d)-[r:DDI {Level:'Major'}]-() OR (d)-[r:DDSI]-() RETURN p
```

Drug	RXCUI	IntDrug	Formula	Description	ATC
"Metformin"	"39998"	"Zonisamide"	"C8H8N2O3S"	"Zonisamide is a sulfonamide anticonvulsant used to treat partial seizures."	"N03AX15; G01AE10;"
"Metformin"	"6826"	"Methazolamide"	"C5H8N4O3S2"	"Methazolamide is a carbonic anhydrase inhibitor used to treat open angle glaucoma and acute angle closure glaucoma."	"G01AE10; S01EC05;"
"Metformin"	"27793"	"Ioxilan"	"C18H24I3N3O8"	"Ioxilan is a diagnostic contrast agent used in various medical imaging procedures; such as angiography; urography; and computed tomographic scans."	"V08AB12"
"Metformin"	"27792"	"Ioversol"	"C18H24I3N3O9"	"Ioversol is a diagnostic contrast agent used in various medical imaging procedures; such as angiography; urography; and computed tomographic scans."	"V08AB07"
"Metformin"	"1546451"	"Iothalamic acid"	"C11H9I3N2O4"	"Iothalamic acid is a diagnostic contrast agent used in various medical imaging procedures;"	"V08AA04"

Fig. 2. The 'Major' adverse reactions of Metformin drug.

In this Cypher query, we used both DDI and DDSI edge properties indicating drug-drug-interaction and drug-disease interactions respectively. The DDI is grouped into three 'Level' categories like 'Major', 'Moderate', and 'Minor' conflict. The result returned 91 nodes that include 15 drugs and 76 UMLS nodes. There are 97 edges in this graph that comprises of 14 DDI edges and 83 DDSI edges. Figure 2 shows the query of Fig. 1 with little modification. Figure 1 is primarily for human consumption, whereas Fig. 2 presents the result that will be ingested by an application through API. Figure 1 enquiry was for both DDI and DDSI, whereas Fig. 2 contains only the DDI. The results here include drug names, drug RXCUI (RxNORM) code, the chemical formula of the

reacting medicine, description of the reacting drug, and the ATC (Anatomical Therapeutic Code) code. The ATC code is used to identify an alternate medicine that will not
react with metformin. The cypher query API used in Fig. 2 query is as follows:

```
MATCH       p=(d:Drug       {RXNORM:'RXNORM:6809'})-[r:DDI
{Level:'Major'}]-(D:Drug) RETURN d.Name AS Drug,D.RXCUI AS
RXCUI,D.Name AS IntDrug,D.MolFormula AS Formula,D.Desc AS
Description,D.ATC AS ATC
```

5 Drugomics Use Case with Clinical Decision Support

Here we present the AI-driven Evidence-Based Clinical Decision Support (AIdEB-CDS)
system that covers a patient's journey from diagnosis to therapeutics. This system is
currently under ethical clearance for point-of-care use at a leading teaching hospital in
Bangalore, India.

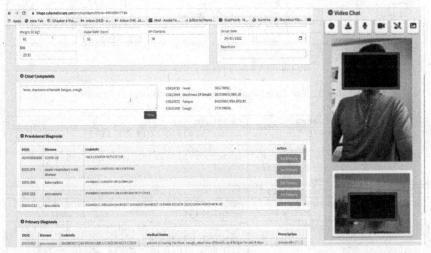

Fig. 3. The Workbench of the Artificial Intelligence driven Evidence Based Clinical Decision
Support (AIdEB-CDS).

There are many rule-based CDS, or Diagnostic-AI systems built since 1970s [20].
Some of these CDS (expert) systems are also known as 'symptom-checkers' as offered by
many healthcare providers [21]. For multiple co-occurring symptoms or for uncommon
symptoms, rule based CDS often fail. Unlike other CDS systems that deal with only the
diagnosis part of the clinical decision, AIdEB-CDS covers the entire patients' journey
from accurate diagnosis to safe therapeutics. Medical error is the third leading cause of
death in the US [22]. Because, AIdEB-CDS ensures safe medical interventions, it will
eliminate medical errors and drug conflicts. Figure 3 depicts the doctor's workbench
AIdEB-CDS application that accesses the knowledge graphs at the Point-of-Care. It

shows the integration of symptomatics, diseasomics, drugomics, and WebRTC. These phases are described below.

5.1 Chief Complaints

The caregiver enters the signs and symptoms of the patient in the 'Chief Complaint' section of the application. These signs and symptoms are presented to the AI-driven evidence-based symptomatomics. In the example (Fig. 3) Chief Complaints section, the doctor entered four symptoms namely 'fever', 'shortness of breath', 'fatigue', and 'cough'. The AI engine uses NLP to extract the medical n-gram terms. The AI engine converts these human-understandable symptoms into machine-understandable UMLS, SNOMED, and ICD10 codes (see the right-hand frame of chief complaints in Fig. 3) as presented in Table 1.

Table 1. The Symptomatomics Knowledge Graph results in Chief Complaints

UMLS	Symptom	SNOMED/ICD10
C0424755	Fever	50177009/_
C0013404	Shortness Of Breath	267036007/R06.00
C0015672	Fatigue	84229001/R54,R53.83
C0010200	Cough	272039006/_

Table 2. The Diseasomics output in Provisional Diagnosis

DOID	Disease	CodeInfo
DOID:0080600	COVID-19	UMLS:C5203670 NCIT:C171133
DOID:974	upper respiratory tract disease	SNOMEDCT:195823002 UMLS:C0029581
DOID:399	tuberculosis	SNOMEDCT:15202009 UMLS:C0041295
DOID:552	pneumonia	SNOMEDCT:266391003 UMLS:C0032285 NCIT:C3333
DOID:6132	bronchitis	SNOMEDCT:155512004;SNOMEDCT:155566007;SNOMEDCT:32398004 ICD10CM:J20;ICD10CM:J40;ICD10CM:J42 UMLS:C0006277;UMLS:C0008677;UMLS:C0149514 NCIT:C26722;NCIT:C26932;NCIT:C2911
etc	etc	etc

5.2 Provisional Diagnosis

The machine interpretable symptoms are used to fetch the likely diseases from diseasomics knowledge graph. It fetched 22 diseases from the diseasomics. The top five results of the query are presented in Table 2 and is also shown in the Provisional Diagnosis section of Fig. 3,

5.3 Prescription

The physician selects a disease as the primary disease from the provisional diagnosis table. The selection of the primary disease opens the prescription section of the application (not shown). The physician writes the medical note and prescribes the medicines. The NLP engine takes the medicine name, normalizes it, and then fetches the RxNORM code for this medicine. All these are then presented in the dashboard in the Primary Diagnosis section (Fig. 3).

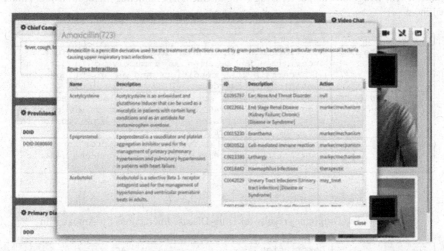

Fig. 4: The prescription of Amoxicillin triggers the drug-drug interaction and the drug-disease interaction to alert the doctor of potential adverse effects (Table 3).

Table 3. Primary Diagnosis with Prescription

DOID	Disease	CodeInfo	Prescription
DOID:552	pneumonia	SNOMEDCT:266391003 UMLS:C0032285 NCIT:C3333	amoxicillin (723)

5.4 Primary Diagnosis

Amoxicillin is shown in primary diagnosis with RxNORM code 723 (see Primary Diagnosis). The physician clicks on the RxNORM code 723 to obtain the information of DDI and DDSI for this drug (amoxicillin in this case). This is shown in Fig. 4. The results are presented in Table 4 and Table 5.

5.5 Drugomics Interactions

The drugomics interface in Fig. 4 shows the drug interactions of amoxicillin. This is very similar to what we have seen in Fig. 1 and Fig. 2. The left panel of the application relates to the DDI (see Table 4). The right panel shows the DDSI. It includes both drug indications and contraindications (see Table 5). The 'therapeutic' value in action column indicates drug-indication, which means that this drug can be used for this disease. Whereas the 'marker/mechanism' in the action column indicates drug contraindication, which means that this drug should not be used if the patient is also suffering from the disease mentioned here.

Table 4. The Drugomics output for Drug-Drug Interaction (Conflicting Medications) for Amoxicillin

Name	Description
Acetylcysteine	Acetylcysteine is an antioxidant and glutathione inducer that can be used as a mucolytic in patients with certain lung conditions and as an antidote for acetaminophen overdose.
Epoprostenol	Epoprostenol is a vasodilator and platelet aggregation inhibitor used for the management of primary pulmonary hypertension and pulmonary hypertension in patients with heart failure.
Acebutolol	Acebutolol is a selective Beta 1- receptor antagonist used for the management of hypertension and ventricular premature beats in adults.

Table 5. The Drugomics results for Drug-Disease Interactions for Amoxicillin

ID	Description	Action
C0395797	Ear; Nose And Throat Disorder	null
C0022661	End Stage Renal Disease (Kidney Failure; Chronic) [Disease or Syndrome]	marker/mechanism
C0015230	Exanthema	marker/mechanism
C0020522	Cell-mediated immune reaction	marker/mechanism
C0023380	Lethargy	marker/mechanism
C0018482	Haemophilus Infections	Therapeutic
C0042029	Urinary Tract Infections (Urinary tract infection) [Disease or Syndrome]	May treat

6 Conclusion

Drug side-effect is a major cause of morbidity and mortality across the world. The root cause of this patient harm is the knowledge and inferential gaps at the point-of-care. We constructed the drugomics knowledge graphs to overcome these constraints during the prescription of medicines. An important use case of the drugomics knowledge graph will be to check if there are any conflicts in medication plans. As a use case, we presented AI-driven Evidence-Based Clinical Decision Support (AIdEB-CDS) system that cover the patients' journey from diagnosis to therapeutics. Such a system could also be integrated with any EHR (Electronic Health Record) system through the REST API to eliminate patient harm. This actionable machine-interpretable knowledge will help the clinician to prescribe the right medication. All these knowledge graphs function like a coordinated 'Physicians' Brain Digital Twin'.

We used digital health – therefore all interactions are both human-understandable and machine-interpretable. We used NLP and other AI functions for clinical and medical documentation in an autonomous way. We used controlled vocabularies like UMLS, ICD10, SNOMED, NCIT, RxNORM. These vocabularies make this system interoperable. This will reduce the documentation time and make more time available for the patient care. Large scale studies are required to be carried out to quantify the effectiveness of such systems.

References

1. Preventable Adverse Drug Reactions: a focus on drug interactions. https://www.fda.gov/drugs/drug-interactions-labeling/preventable-adverse-drug-reactionsfocus-drug-interactions
2. Medication without Harm. https://www.who.int/initiatives/medication-without-harm
3. Tariq, R.A., Vashisht, R., Sinha, A., et al.: Medication dispensing errors and prevention. [Updated 2022 Apr 4]. In: StatPearls [Internet]. Treasure Island (FL): StatPearls Publishing; (2022). https://www.ncbi.nlm.nih.gov/books/NBK519065/
4. Talukder, A.K., Selg, E., Haas, R.E.: Physicians' Brain Digital Twin: Holistic Clinical & Biomedical Knowledge Graphs for Patient Safety and Value-Based Care to Prevent The Post-pandemic Healthcare Ecosystem Crisis. In: Villazón-Terrazas, B., Ortiz-Rodriguez, F., Tiwari, S., Sicilia, M.A., Martín-Moncunill, D. (eds.) Knowledge Graph and Semantic Web, vol. 1686, pp. 32–46. Springer, Cham (2022). https://doi.org/10.1007/978-3-031-21422-6_3
5. Talukder, A.K., Schriml, L., Ghosh A., Biswas R., Chakrabarti P., Haas, R.E.: Diseasomics: Actionable Machine Interpretable Disease Knowledge at the Point-of-Care. PloS Digital Health. Accepted for publication (2022)
6. Building the foundation for better health outcomes. https://go.drugbank.com/
7. Wang, M., Wang, H., Liu, X., Ma, X., Wang, B.: Drug-drug interaction predictions via knowledge graph and text embedding: instrument validation study. JMIR Med Inform. 9(6), e28277 (2021). https://doi.org/10.2196/28277. PMID: 34185011; PMCID: PMC8277366
8. Pryor, T.A., Dupont, R., Clay, J.: A MLM based order entry system: the use of knowledge in a traditional his application In: SCAM, pp. 573–583 (1990). https://europepmc.org/backend/ptpmcrender.fcgi?accid=PMC2245541&blobtype=pdf
9. Drugs@FDA: FDA-Approved Drugs. https://www.accessdata.fda.gov/scripts/cder/daf/
10. Stanford Network Analysis Project. https://snap.stanford.edu/biodata/datasets/10001/10001-ChCh-Miner.html
11. Tatonetti, N.P., Ye, P.P., Daneshjou, R., Altman, R.B.: Data-driven prediction of drug effects and interactions. Sci. Transl. Med. 4(125), 125ra31 (2012). https://doi.org/10.1126/scitranslmed.3003377
12. Xiong, G., et al.: DDInter: an online drug–drug interaction database towards improving clinical decision-making and patient safety, Nucleic Acids Res. 50(D1), D1200–D1207 (2022). https://doi.org/10.1093/nar/gkab880

13. Celebi, R., Uyar, H., Yasar, E., Gumus, O., Dikenelli, O., Dumontier, M.: Evaluation of knowledge graph embedding approaches for drug-drug interaction prediction in realistic settings. BMC Bioinf. **20**, 726 (2019). https://doi.org/10.1186/s12859-019-3284-5
14. FDA Adverse Event Reporting System (FAERS): Latest Quarterly Data Files. https://www.fda.gov/drugs/questions-and-answers-fdas-adverse-event-reporting-system-faers/fda-adverse-event-reporting-system-faers-latest-quarterly-data-files
15. RxNORM. https://www.nlm.nih.gov/research/umls/rxnorm/index.html
16. van Tongeren, J.M.Z., Harkes-Idzinga, S.F., van der Sijs, H., Atiqi, R., et al.: The development of practice recommendations for drug-disease interactions by literature review and expert opinion. Front. Pharmacol. **11**, 707 (2020). https://doi.org/10.3389/fphar.2020.00707
17. DAILYMED. https://dailymed.nlm.nih.gov/dailymed/
18. Sharp, M.E.: Toward a comprehensive drug ontology: extraction of drug-indication relations from diverse information sources. J. Biomed. Semant. **8**(1), 2 (2017). https://doi.org/10.1186/s13326-016-0110-0
19. Neo4j Graph Database. https://neo4j.com/docs/
20. Chen, J.H., Dhaliwal, G., Yang, D.: Decoding artificial intelligence to achieve diagnostic excellence: learning from experts, examples, and experience. JAMA **328**, 709–710 (2022). https://doi.org/10.1001/jama.2022.13735
21. Mayo Clinic Symptom Checker. https://www.mayoclinic.org/symptom-checker/select-symptom/itt-20009075
22. Makary, M A., Daniel, M.: Medical error—the third leading cause of death in the US. BMJ, **353**, i2139 (2016).https://doi.org/10.1136/bmj.i2139

Extremely Randomized Tree Based Sentiment Polarity Classification on Online Product Reviews

R. B. Saranya[1]([✉]), Ramesh Kesavan[2] [iD], and K. Nisha Devi[3]

[1] Narayanaguru College of Engineering, Kanyakumari, Tamil Nadu, India
rbsaranyaravi@gmail.com
[2] Anna University Regional Campus, Tirunelveli, Tamil Nadu, India
[3] Bannari Amman Institute of Technology, Sathyamangalam, Tamil Nadu, India

Abstract. The sentiment analysis of user reviews on social networking is one among the fundamental process carried out by online business organizations to improve the quality of their products and retain the customers and thereby lift the monetary benefit. Although numerous analysis models exist, still there is space for improving the performance and accuracy of informal text data-based classification models. In this examination, we conduct a comparative assessment of the effectiveness of unigram feature set extricated utilizing n-gram technique with three ensemble methods namely Extremely Randomized Tree, Voting, and Bagging classifier based on the following five baseline classifiers Random Forest, Naïve Bayes, K-NearestNeighbor, Ridge Classifier, and Support Vector Machine to identify polarity from mobile product reviews. Among the three ensemble methods, the Extremely Randomized Tree technique has better outcomes with the accuracy of 98% for positive and 85% for negative cases, with an overall accuracy of 96.8%. The error rate of all the three ensemble classifiers is also under 0.5% which uncovers that ensemble classifiers performs better compared to individual classifiers.

Keywords: Ensemble classifiers · Online reviews · Base classifiers · Polarities

1 Introduction

User sentiment have an unprecedented reach to the globe through Internet sources in the current era of social media developments [1]. As per Joseph Johnsons study on global digital population (January 2021) around 60% of the world population is associated through internet (4.66 billion active internet user) out of which 91% of internet users get connected through mobile devices and furthermore there are 4.14 billion active social media users. The survey also states that an internet user spends approximately 8 hours per day on digital media [2]. Quick development of digital device users' produces sheer volume of user generated content which provides potential information and knowledge to governments, businesses, and users themselves [3]. Knowledge extracted from user

© The Author(s), under exclusive license to Springer Nature Switzerland AG 2022
P. P. Roy et al. (Eds.): BDA 2022 India, LNCS 13773, pp. 159–171, 2022.
https://doi.org/10.1007/978-3-031-24094-2_11

generated content support in critical decision making for both consumers and organizations [4–7]. For suggestions on purchase, customers vigorously depend on the feedback available in the e-commerce sites, blogs, news forums, tweets and so on. [2, 4]. To obtain the overall view and opinion of a product or service one must read all the reviews related to it [2, 5]. It is a herculean task for an individual to check and sort enormous number of documents manually. Consequently, it is a fundamental alternative to build automated tools and techniques which will assist users with getting the ideal information from the assortment of reviews [2].

The customers sentiment and attitude posted as feedback or reviews can be interpreted and categorized into positive, negative and neutral polarity.Moreover, Data analytics provide different techniques for the active processing of the above mentioned categorization of the mass amount of online data [8]. A Popular method called Sentiment Analysis (SA), otherwise called emotional polarity computation, is a multidisciplinary part of science that deals with the study, analysis and categorization of the polarity of people's opinions or emotions and attitudes towards various entities [3, 5]. SA is used in many areas other than business process models, including education, e-commerce, politics, disaster management, health, security, and so on [7]. There are myriad techniques like data mining, information retrieval algorithms, machine learning and natural language processing for sentiment analysis [9–11].

According to the mechanism of Sentiment Analysis, classification can be of three levels; in particular, Document level, Sentence level, and Aspect level or Feature level. At document level, the goal is to track down the overall opinion of the entire document.At sentence level, the objective is to identify sentiment orientation in each sentence.In Aspect-level analysis, the characteristics of an object commented by the user are first recognized and thereafter targeted sentiment is determined [4–6, 12].

Machine Learning and Lexicon-based approaches are utilized to determine the sentiment of a text document. Some of the commonly used machine learning based classifiers are Naïve Bayes (NB), Support Vector Machine (SVM), K-Nearest Neighbor (KNN), Logistic regression, Stochastic Gradient Descent, Maximum Entropy, and Random Forest (RF) classifier [13, 14]. Lexicon-based approach is pattern based which identifies the rundown of words and expressions that comprises of semantic worth [1, 12].

The techniques used for sentiment classification on the social media content are categorized as supervised, unsupervised, semi supervised learning, and ensemble techniques [7]. Ensemble classification learning produces a bunch of base classifiers utilizing distinctdisseminations of training data and then classifies by aggregating their outputs. These ensemble learning methods enable users toaccomplish more exact forecast by combining a collection of ensemble base learners, thereby reducing bias and variance with respect to a single base learner. Ensemble techniques higher generalization abilities enhance the predictions performance when compared with individual model-based classification [15]. Remarkable ensemble methods include Bagging, Random Forest, Voting, Boosting, Conditional Forest, and so forth. To train the basic model the Bagging method randomly samples the training data set, to generate sample subsets; which is carried out in a parallel manner. There is an ensemble technique called Voting Classifier (VC) which joins various base classifiers to do classifications through comprising, soft and hard voting schemes [16]. The randomized tree classifier performs the classification task by tuning a weak classification trees with a predefined number of randomly chosen features to create a solid classifier [17].

Data heterogeneity and multi dimensionality are the prime factors that make social media data processing and analysis difficult. Notwithstanding the abovementioned challenges the different difficulties that emerges are domain dependence, negation handling, opinionated phrases, and expressions.Bag-of-words (BoW), higher-order n-grams, POS dependent features, word pairs and dependency relations are some of the feature extraction methods that have been used to improve the classification [17]. Once the sentiment words or phrases have been defined for a specific domain, it can be analysed for their sentiment polarity [17, 18]. This study focuses on sentence-level classification. To exemplify this fact, Twitter tweets on online product are considered for the present study. Twitter is one among the fast-growing microblog where individuals share their perspectives and assessments about a particular point. The main challenges of Twitter tweets are it is generally written in the form of informal languages, short messages that show limited tip about sentiments, and also abbreviations which are widely used [19].

The fundamental commitment of this paper is as per the following:

- The study, start with pre-processing of text data, followed by feature extraction using unigram technique.
- Next, this study identifies the best classification technique among the three ensemble classifiers (Extremely Randomized Tree, Voting, and Bagging classifier) and five base line classifiers (RF, NB, KNN, Ridge Classifier and SVM).

The layout of the paper is structured as follows: The related work in the field of sentiment analysis and ensemble learning is presented in Sect. 2. Section 3 explains the detailed architecture and methodology. Section 4 describes the proposed classification algorithm. The particulars about experiments outcomes are explained in Sect. 5. Finally, Sect. 6 concludes with a discussion of the proposed method as well as with suggestions for future steps.

2   Related Works

Sentiment analysis is used in wide range of research areas and organizations [20]. Numerous researchers build user sentiment classification system to mine the business value from reviews to improve the marketing analysis, general assessment of public, and most importantly customer satisfaction. There has been innumerable research carried on using various feature extraction techniques and classifiers in sentiment analysis. Following is a portion of the noteworthy investigations:

Many researchers have performed a comprehensive study by comparing the performance of ensemble classifier and base line classifiers. Trial discoveries uncover that the performance of ensemble classifiers is better than individual classifiers. The ensemble classifier namely Majority Voting, RF, Bagging Classifier, Ada Boost, Gradient Boost method, Extra Tree, XGBoost, and Stacking are the most popular classifiers used in the recent studies. The most prominent base classifiers used in the latest studies are RF,

SVM, NB, Decision Tree,K-NN, Maximum Entropy, Artificial Neural Network, Logistic Regression along with TF-IDF, n-gram feature extraction technique [1, 4, 15, 19, 21–24].

According to this paper [6] an ensemble method (Voting) for SAusing POS and n-gram feature extraction technique by considering sentiment clue, semantics, and order between words called EnSWF. The result shows that EnSWF out-performs the existing techniques. Similarly, authors [16] in performed Sentiment analysis on US airline twitter dataset using TF-IDF and word2vec features. The exploratory outcome uncovers that the performance of Voting Classifier is superior to other Machine Learning classifiers.

As in [20] a hybrid model which combine Gradient Boosting and Support Vector Machine. It is assessed utilizing two datasets with TF and TF-IDF including unigram, bigram, and trigram feature sets. The study reveals that GBSVM gives preferable outcome over all the other methods by means of performance.

In paper [7] a hybrid ensemble learning model which comprises Ada Boost with SMO-SVM and Logistic Regression in twitter dataset. In this investigation likewise, hybrid ensemble learning model gave preferred outcome over stand-alone classifier (Table 1).

Table 1. Summary of existing work on sentiment analysis using ensemble classifiers.

Author	Feature	Classifier	Dataset	Classification
[15]	TF-IDF with Unigram, Bigram	RF, ET, Bagging, Ada Boost, and Gradient Boost Method	Movie reviews	Binary classification
Ghosh et al. [4]	Unigram, Bigram	KNN, SVM, NB, Maximum Entropy	IMDB Movie Reviews and Product reviews	Binary classification
Sangam and Shinde [24]	Unigram, Bigram	SVM, Artificial Neural Network	Movie reviews	Binary classification
Saleena [19]	BOW	NB, SVM, Logistic Regression, Random Forest	Twitter reviews	Binary classification
Hakak et al. [21]	Number of words, characters, sentences, average word length	Decision Tree, RF, ET	Liar and ISOT reviews	Binary classification
Khan et al. [6]	POS tagging, N-gram	SVM, NB, Generalized Linear method	Movie reviews and Product reviews	Binary classification

(*continued*)

Table 1. (*continued*)

Author	Feature	Classifier	Dataset	Classification
Khalid et al. [20]	TF-IDF, Unigram, Bigram	Gradient Boosting, SVM	Twitter dataset	Binary classification
Sharma and Jain [7]	TF-IDF, N-gram	Ada Boost with SMO-SVM, Logistic Regression	Twitter dataset	Multi classification
Rustam et al. [16]	TF-IDF, Word2vec	SVM, NB, DT, ET, RF, Logistic Regression, Stochastic gradient descent classifier, Calibrated classifier, Ada Boost, Gradient Boosting Machine	Twitter dataset	Multi classification

Review of related literatures interpret that, despite the availability of numerous SA models, still there are opportunities for enhancement of prediction accuracy. In machine learning models the performance can be improved by including pre-processing strategy and specific feature selection strategy. Further Machine Learning models and Ensemble classification plays an essential part in SA.

3   Methodology

The current research devises three ensemble classification models with five baseline classifiers and recommends the best classification model for online product review. Figure 1 shows the experimental procedure.

3.1   Data Set

In this experiment, mobile product review datasets are considered for this analysis is collected from Flipkart, an online e-commerce platform. Totally 8,831 reviews were considered out of which 7,999 reviews are positive cases and 831 are negative reviews. Among the total 8831 reviews, 6185 (70%) reviews are used to train the model, and to access the classification efficiency the model is deployed with 2,646 (30%) independent test datasets out of which 2,401 datasets are positive comments and 245 data are negative comments. Table 2 envisions few reviews statements posted by users to represent positivity and negativity on the product.

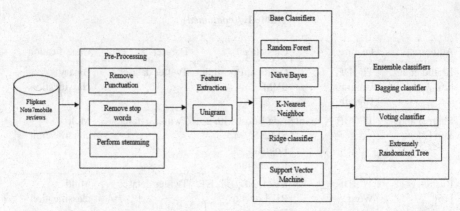

Fig.1. Ensemble classifier-based sentiment polarity classification process on mobile reviews

Table 2. Sample user reviews and their sentiment polarities of mobile products.

Positive polarity	Negative polarity
Nice camera nice product nice performance nice gaming battery backup very good	It's bad mobile. heating problem. hanging problem. many times, display not respond. memory card reading shows in a gallery after Very long time (48 h min.) speaker voice is bad on talking time. overall, it is bad mobile
First of all, the Look and feel of mobile looks good. It feels as a premium phone. Battery of phone becomes decent due to memory optimisation. Display is good as well. on the other hand, call quality is good	Poor phone. Not buy this phone. camera 12 mp.but company say 48 mp.very bad experience
Best phone in this price range. The camera quality is amazing. battery backup is also nice. Overall, very good product	Worst product. no proper working face lock not working battery very fast drain no one to buy its waste of money no one response for my complaint
Great Mobile. Excellent Rear Camera. Photos are sharp and clear. 48 MP camera quality is too good even in the dark place or at night time. Front Camera is also very good. Stylish Design and Glassy Finish of backside gives Premium Look. Good performance for 6 GB RAM Variant. Type C charging port with 4.0 quick charging support	Don't Buy this mobile very bad - Horrible Battery performance. Drained immediately. And customer support is pathetic

3.2 Text Pre-processing

Data quality improvement has a significant influence in the classification outcome. To improve the quality the following pre-processing techniques namely tokenization, punctuation removal, stemming, and stop words removal were applied on the experimental dataset.

3.3 Feature Extraction

Feature extraction in text analysis is the way of determining a bunch of discriminative, instructive and non- redundant numerical values. In this study, the feature extraction is done using unigram technique a variant of n- gram technique.

3.4 Unigram Model

Unigram language model techniqueismostly used for text categorization because of their language independent nature. Unigram model throws all single contexts in the sentence and estimate the probability of each word independently.

$$P_{unigram}(w_1 w_2 \dots . w_n) = P(w_1)P(w_2) \dots . P(w_n)$$

Where w_i is the individual term in the sentence.

4 Classification

There are two types of Sentiment classification: binary sentiment classification and multiclass sentiment classification [25]. In this study, implementation is done with respect to binary classification to classify customers opinion on a mobile product as positive or negative [26].

4.1 Ensemble Methods

Ensemble classifier derives a new classifier by combining base classifiers to improve the classification accuracy.

Voting Classifier. The Voting Classifier (VC) is an ensemble technique which joins various base models to do classifications through comprising, soft and hard voting schemes [7]. In this paper, we adopted hard voting. The class label with the highest votes will be chosen in a hard voting. The classification learners RF, NB, KNN, RC, and SVM were used as base classifier. The validation is carried out using 10- fold cross validation method. The class label $\hat{y}$ is predicted by calculating the mode of all classifier C_j. Where j varies from 1 to m and X is the number of features [27].

$$\hat{y} = mode\{ \ C_1(X), \ C_2(X), \ \dots, \ C_m(X)\} \hat{y} = mode\{ \ C_1(X), \ C_2(X), \ \dots, \ C_m(X)\}$$

The class label $\hat{y}$ is predicted by calculating the mode of all classifier C_j Where j varies from 1 to m and X is the number of features [27].

Bagging Classifier. Bagging classifiers are ensemble meta-estimators that use the best mean probability or major voting method to construct classification models. To train the basic model it randomly sample the training data set, which is independent of other sample [23]. As in voting classifier k fold validations are done in bagging classifier. In our study the parameter value tuned to ensemble bagging classifiers are random_state is set to seed, max_samples as 0.4 and max_features are fixed to 10.

Randomized Tree Classifier. Randomized tree Classifier comprises numerous different sets of classifiers which is used to build a group of decision trees. While partitioning a tree, this approach produces totally randomized trees with structures that are independent of the output values [28, 29]. We set the value of n_estimator to 300 in this classification to increase the predictive accuracy and control over- fitting.

4.2 Base Classifiers

In ensemble methods, learners composing an ensemble are normally called base learners. The input given for ensemble classifiers is RF, NB, KNN, RC and SVM.

Random Forest. Random Forest is an ensemble learning classification which is comprised of numerous different sets of classifiers which is used to build a group of decision trees. This method can be implemented by giving a test sample to the new classifier and the sample class label can be determined based on the voting result of each classification [23]. Ultimate conclusion depends on the maximum voted class in the forest and the class with highest vote is considered to be the output [30, 31]. Tuning of hyper parameters in random forest needs a special attention since they are needed to be tuned manually so it is extremely tedious and complex. In this study, RF classification was set with the number of features at random 0, depth value was set to unlimited, n_ estimators were examined and evaluated with ranges {100, 200, 500, 1000} in which n_estimators with 1000 gave an optimal result. These values were picked based on the most elevated outcome.

K-Nearest Neighbor. The KNN classifier determines the labels for unknown sample based on the closeness to the training samples, where K is the number of classes.The bootstrap procedure was used to calculate the parameter K [32]. To determine the optimal K value, we examined the range of K values from 1 to 10.

Naïve Bayes. The Naïve Bayes classifier is a Bayes' theorem-based classifier. There are different types of Naïve Bayes classifier among them Multinomial Naive Bayes classifier is used in this study which is most preferable for text classification [11].

Ridge Classifier. Ridge regression is a variant of linear regression which avoids the complexity and over-fitting of simple linear regression. The Ridge regressor has a classifier variant called Ridge classifier.To overcome the overfitting problem,ridge regression adds a hyperparameter λ a penalty term value ranging from 0.1 to 1.0. In this study the linear model was chosen; meanwhile the λ value is set to 0.5 respectively.

Support Vector Machine. In SVM classification process each data item is plotted as a point in n-dimensional space, with the values of each feature being the value of certain coordinate. Hyper-planes are used to analyse data and set decision limits in this method. In this worklinear Kernel is utilized with a learning rate of 0.1, to obtain reliable results on categorical data [25, 33].

4.3 Performance Evaluation Parameters

In view of the values acquired from confusion matrix the performance of individual classifiers is assessed utilizing the parameter's precision, recall, F-measure, and accuracy [15, 16, 20] (Table 3).

Table 3. Evaluation parameters.

Evaluation parameter	Description	Formula
Precision	It measures the exactness of the classifier and involves TP to the sum of TP and FP	Positive Precision = TP/(TP + FP) Negative Precision = TN/(TN + FN)
Recall	It measures the completeness of a classifier	Positive Recall = TP/(TP + FN) Negative Recall = TN/(TN + FP)
F1-score	It defines the harmonic mean of exactness and completeness	F1-Score = (2*Precision*Recall)/ (Precision + Recall)
Accuracy	It defines the ratio of correctly classified data to total amount of data	Accuracy = (TP + TN)/(TP + TN + FP + FN)

5 Result and Discussion

This part contains the subtleties of the experimental results along with the interpretation obtained from the outcomes. To categorize sentiments into positive and negative classification, experiments are done utilizing five basic classifiers and three ensemble classifiers. Based on the elements of confusion matrix the performance is evaluated.

Table 4 presents the performance analysis of individual learners and ensemble learners. Extremely Randomized Tree ensemble classifiers achieves the best performance with an overall accuracy of 96.8%, which also has the highest F-measure value 98% for positive comments and 83% for negative comments compared to the rest of the classifiers implemented in this research. The True Positive (TP) and True Negative (TN) values obtained are 98% and 85% respectively which shows that Extremely Randomized tree shows better performance in both positive and negative comments when compared to all other classifiers. The error rate is 0.031 which is also very low when compared to other classifiers. On comparing with the Extremely Randomized Tree classifier the Bagging and Voting classifiers has a slight degradation in the performance with 95.7% and

94.5% accuracy respectively. Assessment on the recall and F-measure reveals that all the three-ensemble model's performance are the same for positive polarity classification but in the case of negative polarity the accuracy of Extremely Randomized tree is better with 85% against 83% of bagging and 80% of voting classifier model.

Among the base classifier's RF and SVM are close challengers with accuracy 93.9% and 93.5%, though the accuracy of SVM is higher than RF, the precision value is lower in positive comments and higher in negative comments. All performance indices that sensitivity of SVM classifier is higher when compared to the rest of the classifiers implemented in the current research. Although the TP value for RF classifier is 99% which is highest among all the classifiers, the False Positive (FP) count around 56% states that the RF classification model is skewed towards positivity. However, the KNN classifier which is second among all the individual classifiers which achieved an accuracy of 92.8%, which gives 96% performance over all evaluation measures in positive comments and 61% in negative comments. NB and Ridge classifier achieves the accuracy of 90.1% and 90.3% which is almost same.

The overall performance result acquired from the unigram vectorization model with five base learners and three ensemble learners are summarized in the Fig. 2. The true positive measure which represents the number of predicted cases matching with the actual cases. Study on true positive cases conveys that RF, KNN and all three ensemble classifiers have done a better classification on positive cases. Interpretation from true negative cases reveals that other than RF, NB and KNN all other classifiers have given more or less a even number of right classification of negative comments. Though this classification model is on mobile review whose outcome is not vital, generally False Positive prediction has a big concern over False Negatives prediction. Though SVM model has given a better negative comment classification false positive alarm makes this model unsuccessful. The high count of false negative classification of RF model leads to the low categorization of true negative cases and thus degrades the RF model.

Table 4. Performance analysis of RF, NB, K-NN, RC, and SVM classifiers with unigram features

Positive class				Negative class			
Classifier	Precision	Recall	F-measure	Precision	Recall	F-measure	Accuracy
RF	0.99	0.94	0.97	0.44	0.82	0.57	93.9
NB	0.93	0.95	0.94	0.62	0.47	0.54	90.1
KNN	0.96	0.96	0.96	0.61	0.61	0.61	92.8
RC	0.91	0.98	0.94	0.83	0.48	0.61	90.3
SVM	0.94	0.99	0.96	0.89	0.6	0.72	93.5
Bagging	0.97	0.98	0.98	0.83	0.74	0.78	95.7
Voting	0.96	0.98	0.97	0.8	0.67	0.73	94.5
Randomized Tree	0.98	0.98	0.98	0.85	0.81	0.83	96.8

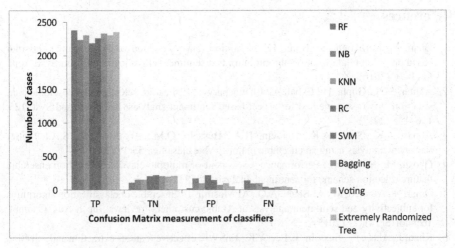

Fig. 2. Summary of confusion matrix based comparative analysis.

Cumulatively, the experimental research reveals that ensemble model's performance is comparatively better than stand-alone models. Among the ensemble models Extremely Randomized Tree model performs a better classification with overall accuracy of 96.8%, which is 1.1% higher than the close competitor bagging model. As human language is intrinsically biased towards positivity, and due to the existence of a higher number of positive comments in the dataset, positive comments are better classified by all base and ensemble classifiers than negative comments. Evaluating negativity in reviews is still a difficult semantic job.

6   Conclusion

A comparative evaluation of three ensemble methods namely Bagging, Voting, and Randomized Tree, and also five baseline classifiers entitled Support Vector Machine, Random Forest, Naïve Bayes, K-Nearest Neighbor, and Ridge Classifier is performed for sentiment classification of mobile product review dataset. Exploratory findings uncover that the ensemble classifiers perform better compared to individual classifiers. Precision, recall, F- measure, and accuracy are used to evaluate the classifier's performance. Among the three ensemble methods, Extremely Randomized tree classifier is evidently identified to produce best results when compared to other classifiers. The accuracy obtained for this classifier is 96.8%. Bagging and Voting has the accuracy of 95.7% and 94.5% respectively. Among the five baseline classifiers RF and SVM classifier are close challengers with accuracy 93.9% and 93.5%. Over the majority of evaluation measures SVM and RF have performed the best. As far as the adequacy with respect to the three ensemble strategies, we may infer that ensemble classification algorithms perform powerfully better compared to the wide range of various individual classifiers. The future scope of this work, will be carried out by relating sentiment classification with product sales promotion for better business intelligence.

References

1. Onan, A., Korukoğlu, S., Bulut, H.: A multiobjective weighted voting ensemble classifier based on differential evolution algorithm for text sentiment classification. Expert Syst. Appl. **62**, 1–16 (2016)
2. Akhtar, M.S., Gupta, D., Ekbal, A., Bhattacharyya, P.: Feature selection and ensemble construction: a two-step method for aspect based sentiment analysis. Knowl.-Based Syst. **125**, 116–135 (2017)
3. Abbas, A.K., Salih, A.K., Hussein, H.A., Hussein, Q.M., Abdulwahhab, S.A.: Twitter sentiment analysis using an ensemble majority vote classifier. **55**(1) (2020)
4. Ghosh, M., Sanyal, G.: Performance assessment of multiple classifiers based on ensemble feature selection scheme for sentiment analysis. **2018** (2018)
5. Omar, N., Albared, M., Al-Shabi, A.Q., Al-Moslmi, T.: Ensemble of classification algorithms for subjectivity and sentiment analysis of Arabic customers' reviews. Int. J. Adv. Comput. Technol. **5**(14), 77 (2013)
6. Khan, J., Alam, A., Hussain, J., Lee, Y.K.: EnSWF: effective features extraction and selection in conjunction with ensemble learning methods for document sentiment classification. Appl. Intell. **49**(8), 3123–3145 (2019)
7. Sharma, S., Jain, A.: Hybrid ensemble learning with feature selection for sentiment classification in social media. **10**(2), 40–58 (2020)
8. Li, N., Wu, D.D.: Using text mining and sentiment analysis for online forums hotspot detection and forecast. Decis. Support Syst. **48**(2), 354–68 (2010)
9. Wang, W., Feng, Y., Dai, W.: Topic analysis of online reviews for two competitive products using latent Dirichlet allocation. Electro. Commer. Res. Appl. **29**, 142–56 (2018)
10. Siering, M., Muntermann, J., Rajagopalan, B.: Explaining and predicting online review helpfulness: the role of content and reviewer-related signals. Decis. Support Syst. **108**, 1–12 (2018)
11. Qi, J., Zhang, Z., Jeon, S., Zhou, Y.: Mining customer requirements from online reviews: a product improvement perspective. Inf. Manag. **53**(8), 951–63 (2016)
12. Da Silva, N.F., Hruschka, E.R., Hruschka, E.R., Jr.: Tweet sentiment analysis with classifier ensembles. Decis. Support Syst. **66**, 170–179 (2014)
13. Dey, A., Jenamani, M., Thakkar, J.J.: Senti-N-Gram: an n-gram lexicon for sentiment analysis. Expert Syst. Appl. **103**, 92–105 (2018)
14. Awwalu, J., Bakar, A.A., Yaakub, M.R.: Hybrid N-gram model using Naïve Bayes for classification of political sentiments on Twitter. Neural Comput. Appl. **31**(12), 9207–9220 (2019)
15. Rahman, S.S.M.M., Rahman, M.H., Sarker, K., Rahman, M.S., Ahsan, N., Sarker, M.M.: Supervised ensemble machine learning aided performance evaluation of sentiment classification. In: Journal of Physics: Conference Series, p. 012036. IOP Publishing (2018)
16. Rustam, F., Ashraf, I., Mehmood, A., Ullah, S., Choi, G.S.: Tweets classification on the base of sentiments for US airline companies. Entropy **21**(11), 1078 (2019)
17. Wang, G., Sun, J., Ma, J., Xu, K., Gu, J.: Sentiment classification: the contribution of ensemble learning. Decis. Support Syst. **57**, 77–93 (2014)
18. Xu, J., Zheng, S., Shi, J., Yao, Y., Xu, B.: Ensemble of feature sets and classification methods for stance detection. In: Lin, Chin-Yew., Xue, Nianwen, Zhao, Dongyan, Huang, Xuanjing, Feng, Yansong (eds.) ICCPOL/NLPCC -2016. LNCS (LNAI), vol. 10102, pp. 679–688. Springer, Cham (2016). https://doi.org/10.1007/978-3-319-50496-4_61
19. Saleena, N.: An ensemble classification system for twitter sentiment analysis. Procedia comput. Sci. **132**, 937–946 (2018)

20. Khalid, M., Ashraf, I., Mehmood, A., Ullah, S., Ahmad, M., Choi, G.S.: GBSVM: sentiment classification from unstructured reviews using ensemble classifier. Appl. Sci. **10**(8), 2788 (2020)

21. Hakak, S., Alazab, M., Khan, S., Gadekallu, T.R., Maddikunta, P.K.R., Khan, W.Z.: An ensemble machine learning approach through effective feature extraction to classify fake news. Future Gener. Comput. Syst. **117**, 47–58 (2021)

22. Li, Y., Chen, W.: A comparative performance assessment of ensemble learning for credit scoring. Mathematics **8**(10), 1756 (2020)

23. Ampomah, E.K., Qin, Z., Nyame, G.: Evaluation of tree-based ensemble machine learning models in predicting stock price direction of movement. Information **11**(6), 332 (2020)

24. Sangam, S., Shinde, S.: Sentiment classification of social media reviews using an ensemble classifier. Indones. J. Electr. Eng. Comput. Sci. **16**(1), 355 (2019)

25. Tripathy, A., Agrawal, A., Rath, S.K.: Classification of sentimental reviews using machine learning techniques. Procedia Comput. Sci. **57**, 821–829 (2015)

26. Tripathy, A., Agrawal, A., Rath, S.K.: Classification of sentiment reviews using N-gram machine learning approach. Expert Syst. Appl. **57**, 117–26 (2016)

27. James, G.: Majority Vote Classifiers: Theory and Applications (1998)

28. Geurts, P., Ernst, D., Wehenkel, L.: Extremely randomized trees. Mach. Learn. **63**(1), 3–42 (2006)

29. Marée, R., Wehenkel, L., Geurts, P.: Extremely randomized trees and random subwindows for image classification, annotation, and retrieval. In: Criminisi, A., Shotton, J. (eds.) Decision Forests for Computer Vision and Medical Image Analysis. Advances in Computer Vision and Pattern Recognition, pp. 125–41 Springer, London (2013). https://doi.org/10.1007/978-1-4471-4929-3_10

30. Onan, A., Korukoğlu, S., Bulut, H.: Ensemble of keyword extraction methods and classifiers in text classification. Expert Syst. Appl. **57**, 232–247 (2016)

31. Parmar, A., Katariya, R., Patel, V.: A review on random forest: an ensemble classifier. In: Hemanth, Jude, Fernando, Xavier, Lafata, Pavel, Baig, Zubair (eds.) ICICI 2018. LNDECT, vol. 26, pp. 758–763. Springer, Cham (2019). https://doi.org/10.1007/978-3-030-03146-6_86

32. Thanh Noi, P., Kappas, M.: Comparison of random forest, k-nearest neighbor, and support vector machine classifiers for land cover classification using Sentinel-2 imagery. Sensors **18**(1), 18 (2018)

33. Winkler, S., Schaller, S., Dorfer, V., Affenzeller, M., Petz, G., Karpowicz, M.: Data-based prediction of sentiments using heterogeneous model ensembles. Soft Comput. **19**(12), 3401–3412 (2015)

Community Detection in Large Directed Graphs

Siqi Chen and Raj Bhatnagar$^{(\boxtimes)}$ (iD)

University of Cincinnati, Cincinnati, OH 45221, USA
`Raj.Bhatnagar@uc.edu`

Abstract. Many real-life systems are represented by directed networks but most of the existing community detection research is focused on undirected networks. We present here a new scalable Map-Reduce algorithm to discover PageRank-driven hierarchical community structures in directed graphs. Main features of our approach are that we seek to find a directed community centered around any selected core node of a directed graph, and a community is defined by two hierarchical structures connected to the core node. An upper structure, that points "to" the core node, and a lower structure that is pointed to "by" the core node. Nodes chosen for inclusion in our hierarchical community structure must display importance ratings (PageRank) very similar to the rating of the selected core node. We have demonstrated here that our algorithm finds meaningful hierarchical communities in real-world data sets. We have also evaluated the effects of the PageRank of the core node and the PageRank threshold parameter for the nodes included in its communities.

Keywords: Directed graphs · Community detection · Graph mining

1 Introduction

A large number of real-world graph data sets are intrinsically directed and directions are an essential aspect of the system. However, most of the existing clustering or community detection approaches have been focused on undirected networks. Edge directionality introduces an extra complication and thus community detection in directed networks is a more challenging task.

Most of the existing work related to community detection in directed graphs suffers from one or more of the following drawbacks: (1) many methods ignore edge directions and assume symmetric interactions; (2) many existing methods are sequential algorithms, that are not easily scalable for very large graphs; (3) most community detection algorithms seek to find strongly connected components of a network and then use them as cores for communities. There is a need to construct communities surrounding individual query nodes of interest that may not be part of strongly connected clusters of a network.

P. P. Roy et al. (Eds.): BDA 2022 India, LNCS 13773, pp. 172–181, 2022.
https://doi.org/10.1007/978-3-031-24094-2_12

Here we present an algorithm to find hierarchical communities of interest in large directed graphs. The challenges that we address relate to two primary aspects of the problem: (1) defining the structure of a meaningful directed community, and (2) developing an efficient scalable Map-Reduce algorithm to find these directed communities in large graphs. The goal is to obtain a hierarchically structured community around a core node such that (1) the core node is located in the center of the hierarchical community; (2) the nodes in upper levels can reach the core node; (3) the nodes in lower levels can be reached from the core node; and (4) the nodes within the community have similar importance (Page Rank) compared to the nodes outside the community.

2   Related Work

One way to handle directed graph clustering is to convert them into undirected ones using symmetrization operations. These transformed graphs can then be clustered using existing methods. Lai et al. [1] proposed a transformation method that uses PageRank induced random walks to obtain a new undirected representation of the original graph. Satuluri et al. [2] discussed random walk based symmetrization, where the directed normalized cut of the original networks is equal to the normalized cut of the transformed undirected network [3]. In addition, the authors also explored several other ways of symmetrizing a directed graph into an undirected one, while information on edge directionality is incorporated into the edge weight of the transformed graph.

The methods discussed above largely benefit from the existing well-developed approaches for community detection in undirected graphs. However, one drawback of these methods is that since edge directionality is considered as an inherent property of the network, any transformation techniques may not be able to completely retain the information on edge directionality. To address this problem, many studies have tended to generalize or extend the clustering methods to directed graphs without changing the structure of the original graph.

PageRank and random walks have been employed in many clustering algorithms for undirected graphs [4–7]. Since they traditionally have been developed and studied in directed web graphs, it is natural to study if they can be generalized to detect clusters in a directed graph. Avrachenkov et al. [8] proposed a PageRank based clustering (PRC) algorithm for hypertext document collections that are represented by directed graphs. This method consists of two steps. The first step is to generate a ranked list of core nodes based on PageRank. Then nodes are assigned to different clusters using their personalized PageRank vectors. In contrast to global clustering of the entire graph, Andersen et al. [9] proposed a local partitioning method that finds a set of nodes near a specified seed node by examining only a small portion of the input directed graph and using personalized PageRank vectors. The authors proved that by sorting the nodes of the graph according to the ratios of the entries in the Personalized PageRank vectors and the global PageRank vectors, this method is able to efficiently detect local clusters.

3   Our Approach

We use the PageRanks of nodes, computed in the global context of the complete graph for determining the importance of a node for belonging to a cluster. For each candidate core node, we extract a forward and a backward directed subgraph to construct the hierarchical community structure surrounding the core node.

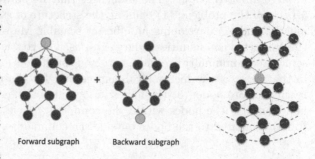

Forward subgraph   Backward subgraph

Fig. 1. A nine-level community around a core node (yellow) by combining five-level forward and backward subgraphs (Color figure online)

3.1   PageRank

PageRank algorithm is commonly used to rank nodes in trust networks and it computes a score for each user that represents the average opinion of all the nodes in the graph about that user.

3.2   Overall Algorithm Outline

We define a directed community around a core node as having the following characteristics:

- A lower subgraph consisting of up to m layers of nodes such that the top node is the core node and each layer has directed edges only to nodes in layers below it. This is shown as the forward subgraph on the left in Fig. 1.
- An upper subgraph consisting of up to m layers of nodes such that a node in each layer has directed edges only to nodes in layers below it, and the lowest layer has only one node, the core node. (Shown as the middle of Fig. 1.)
- Joining the above two subgraphs gives us the $(2 \times m - 1)$ level directed community structure around the core node.
- If the PageRank of the core node is p, we use a PageRank threshold of $k \times p$, where k can be between 0 and 1, and all the nodes contained in all the layers of the community have a PageRank larger than this threshold value. The idea is to form a community in which all nodes are above some minimum threshold of PageRank value. For example, if the PageRank of the core node is 0.25, we may want to include only those other nodes that have a PageRank of 0.2 or higher. This criterion for inclusion of nodes can be adjusted as per the needs of the problem context.
- To determine the intra-community connectivity of a community structured as above, we add back all the edges connecting the nodes of the community from the original complete graph.

The proposed algorithm goes through two stages with three Map-Reduce phases, as shown in Fig. 2. The output of a Map-Reduce phase is chained to the input of its next phase.

- Stage 1: Grow upper-level and lower-level directed subgraphs directed "to" and "from" the core node.
- Stage 2: Combine the resulting upper-level and lower-level subgraphs into a hierarchical community centered at the core node, remove the undesirable nodes, and add all edges connecting the nodes of the remaining community.

Our algorithm outlined below generates the lower-level subgraphs. We repeat the same algorithm by reversing all the edge direction in the input graph to construct the upper-level subgraphs.

Map-Reduce Phase 1: Finding Incoming and Outgoing Nodes. This Map-Reduce phase finds all outgoing and incoming nodes of each node in the graph. The outgoing nodes of a node u can be easily obtained by enumerating all the edges starting from node u. Similarly, the incoming nodes of node u can be obtained

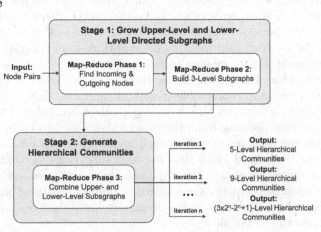

Fig. 2. Flowchart of the three-phase Map-Reduce algorithm

by enumerating all the edges ending at node u. In Mapper1, for each node pair (u, v), two key/value pairs - the original node pair representing the out-link from u and the reversed node pair representing the in-link to u, are emitted, respectively. In Reducer1, the incoming and outgoing nodes of each key node are collected.

Map-Reduce Phase 2: Building Three-Level Subgraphs. From the lists of outgoing and incoming nodes of a node, we generate three-level subgraphs rooted at each incoming node in a Map-Reduce job. In Mapper2, for each key/value pair (key-node, (outgoing-nodes, incoming-nodes)), two-hop paths starting from each incoming node are generated. So in the output key/value pair, the key is a node from the incoming node list and the value is a two-hop path originating from the key node to outgoing nodes. In effect, each Mapper server generates all directed paths of length two starting from each node. In Reducer2, two-hop paths with the same root node are merged into a three-level subgraph. To maintain the hierarchical structure of the community, if node u appears at more than one level in a subgraph, e.g., levels i and j ($i < j$), node u at the deeper level j and its descendants should be removed.

Map-Reduce Phase 3: Building Higher-level Subgraphs. By combining multiple three-level subgraphs through an end-to-end concatenation, a five-level subgraph can be constructed. Similarly, a nine-level subgraph can be created from multiple five-level subgraphs. Ideally, we can construct a subgraph with $(3 \times 2^n - 2^n + 1)$ levels through n iterations. To make this approach scalable, every iteration consists of one Map-Reduce iteration and an additional Reduce step. Here we use the construction of five-level communities as an example to illustrate how this Map-Reduce phase works. In Mapper3, in order to perform the end-to-end concatenation in a key/value pair fashion, the generated three-level subgraph is decomposed into smaller paths by its leaf nodes. A output key/value pair with a leaf node as key and a three-level path from the root node down to the leaf node as value is generated. In Reducer3, these three-level root-to-leaf paths are combined with the original three-level subgraph that share the same key node. Before moving to the final merging step, these branches are further refined based on the PageRank value of the root node. To build a subgraph where all nodes have comparable PageRank values to that of the root node, we can remove those nodes that have PageRank values below a certain value by using a pre-specified PageRank threshold k (between 0 to 1). In Reducer4, all of these five-level branches as well as the original three-level subgraph with the same key node are merged into a final five-level subgraph. An additional branch pruning is required to remove the duplicate nodes and inter-level edges in the merged subgraphs. Using the same pruning approach as stated in the previous section, these duplicate nodes at the deeper level and their descendants are removed to retain the hierarchical structure.

3.3   Time Complexity

Since multiple Mapper servers work in parallel on different chunks of the data, an almost linear scale-up has been observed for this type of enumeration algorithm. For average cases, the complexity has been observed empirically to rise exponentially with the average degree of the network. For the case of the datasets tested by us, a scalability analysis based on the actually observed results is presented in a later section of this paper.

4   Experiments and Results

In this section, we demonstrate how to use our algorithm to build nine-level communities from a real-world directed network, the Bitcoin Alpha dataset. We also investigate the effects of the core node's PageRank score and the PageRank threshold used during the community construction process. We then examine the results of our method using a few evaluation metrics. Our experiments are executed on a cluster of 6 nodes. Each node has 12 CPU cores and 48.25 GB of RAM. The version of Apache Spark employed is 2.3.2. The applications and algorithms are developed in Python and run with PySpark 2.4.3.

4.1 Nine-Level Communities

Bitcoin Alpha dataset is a user-user trust network on a Bitcoin trading platform called Bitcoin Alpha where users can buy and sell products using Bitcoins [10,11]. (https://snap.stanford.edu/data/soc-sign-bitcoin-alpha.html) A node represents a user and a directed edge shows that the left user trusts the right user. In this network, node #4 has the highest PageRank score and thus is the most trustworthy user in the graph, while node #6434 has the lowest PageRank score, indicating that he/she is not trusted by most of the users.

As we can see in Fig. 3, the community discovered around node #4 is small and consists of only 10 other nodes; the community around node #2278 with a medium PageRank score has many more nodes and its upper part and lower part have comparable sizes (subfigure (c)). Node #6434 has s very small PageRank score. It can be clearly seen that in the community of node #6434 (sub-figure (b)), several nodes with much higher PageRank scores than that of node #6434 appear on the second and third levels of the community. These high PageRank nodes attract a lot of deeper-level nodes, which largely increases the size of the community.

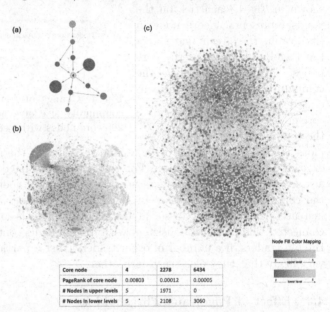

Core node	4	2278	6434
PageRank of core node	0.00803	0.00012	0.00005
# Nodes in upper levels	5	1971	0
# Nodes in lower levels	5	2108	3060

Fig. 3. Bitcoin dataset: nine-level communities of example nodes

4.2 Community Coefficient and Community Size

To measure the quality of communities obtained, we have introduced an evaluation metric called community coefficient (CC). It is the ratio of the actual number of edges in the community to the maximum possible number of edges in the community, defined as: $CC = \frac{\text{\# of actual edges}}{\text{\# of all possible edges}} = \frac{N_{edge}}{C(N_{node},2)\times 2}$, where

N_{edge} is the total number of edges among all the nodes of the generated community in the graph and N_{node} is the total number of nodes in the community. The community coefficient is in the range [0, 1]. The higher the community coefficient, the more densely connected are the nodes in the community.

To find interesting communities, we create communities centered at every possible node in the graph and measure the community size and community coefficient of each community. These two metrics are then plotted for all core nodes in the order of increasing PageRank scores. As shown in Fig. 4, each tick on the x-axis represents a core node in the graph; the green line shows the PageRank scores of the core nodes, the blue line shows the community coefficients of the generated communities around the core nodes, and the red line shows the community sizes. These three plots combined together provide a good insight into the nature of communities existing in the original graph. For the Bitcoin Alpha dataset, as the PageRack score of the core node increases, the community coefficient increases while the community size decreases in general. That is, as the PageRank of nodes increases, the community associated with that person is smaller and more tightly connected. This trend is consistent with the domain's context. In the case of the Bitcoin traders, if a number of other traders trust a particular traders then it is very likely that they also trust each other.

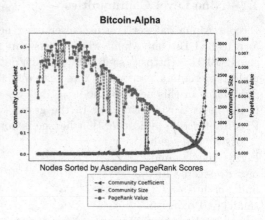

Fig. 4. Change of community coefficient and community size of generated communities for 200 core nodes with decreasing PageRanks

4.3 Effect of PageRank Threshold k

In the community construction process, we select only those nodes that possess comparable or higher importance (PageRank) than that of the core node. Therefore, a PageRank threshold k is applied to all the included nodes in the forward and backward m-level subgraphs rooted at the core node.

We have evaluated the effect of changing the threshold k on the community coefficients of the resulting communities. Figure 5 shows the computed community coefficient values and the community sizes of the resulting communities for top 50 core nodes having the highest PageRanks. The x-axis represents the nodes sorted by their PageRank in descending order. It can be clearly observed that as the PageRank score of the core node decreases, the value of community coefficient drops while the community size increases. When the PageRank score of the core node is high, the size of the generated community is generally small, resulting in a high value of the community coefficient. When the PageRank score of the core node decreases, more nodes with slightly smaller PageRank scores start to be included in the community.

As the community size becomes larger, the cluster connectivity may reduce accordingly, leading to a smaller value of the community coefficient. Meanwhile, with the decrease of the PageRank threshold, the community coefficient decreases and the community size increases. When we lower the PageRank threshold, more nodes with relatively lower PageRank scores are added to the community, which may introduce weak connections to the original community.

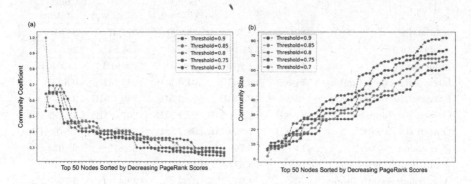

Fig. 5. Bitcoin Alpha dataset: change of (a) community coefficient and (b) community size for top 50 nodes with highest PageRank scores when PageRank threhsold = 0.9, 0.85, 0.8, 0.75 and 0.7

When selecting the PageRank threshold for generating hierarchical communities, it is necessary to consider the community sizes we would like to generate and how large the community coefficients we would like to achieve. Then we can select the PageRank threshold to obtain the communities that meet our desired performance criteria. Here we have shown all the communities generated and their comparison. The system can be easily modified to output only those communities that meet the specified size and community coefficient criteria.

4.4   Scalability Analysis

To evaluate scalability of our approach, we have compared the execution resources needed for our algorithm on the three real-world datasets with different sizes (Physicians, Bitcoin Alpha and Supreme Court Citation) for building nine-level communities when using PageRank threshold $k = 0.8$. To gain further insight into scalability, we have created two smaller versions of the largest dataset, Supreme Court Citation dataset, by randomly sampling one third and two third of its edges. A comparison of the execution resources data including input data size, the volume of the intermediate data in the Hadoop system, output data size, number of generated five-level communities, average community length, and total execution time are shown in Table 1. The intermediate data is the maximum amount of data generated during the Map-Reduce phases. The average community length is the average number of levels in the generated communities. Note that all data processing is carried out using our cluster with 6 Mapper nodes.

Table 1. Execution time and memory needed for constructing nine-level communities.

Parameters	Physicians	Bitcoin-Alpha	Supreme-Court-citation_1	Supreme-Court-citation_2	Supreme-Court-citation
# of nodes	241	3,783	26,720	31,557	34,613
# of edges	1,098	24,186	67,322	134,845	202,167
Input size	54 KB	1.2 MB	3.9 MB	7.8 MB	11.8 MB
Intermediate data (upper)	574 KB	3.2 GB	31.8 MB	233 MB	0.8 GB
Intermediate data (lower)	1,082 KB	2.9 GB	93.3 MB	790.3 MB	2.9 GB
Output size (upper)	164 KB	229.1 MB	29.7 MB	267.2 MB	792. 5 MB
Output size (lower)	247 KB	205.9 MB	98.7 MB	726.5 MB	1,870 MB
# of comm. (upper)	213	3,740	16,939	24,812	29,671
# of comm. (lower)	214	3,273	17,451	21,656	23,512
Avg. comm. length (upper)	4.57	4.95	4.22	4.64	4.79
Avg. comm. length (lower)	4.64	4.92	4.40	4.73	4.8
Total time (upper)	21 s	305 s	37 s	133 s	212 s
Total time (lower)	19 s	298 s	60 s	394 s	634 s

From Table 1, we can see that as the size of the dataset increases, the execution time and the size of the intermediate data increase in most cases. For each dataset, the processing time for generating the upper and lower five-level subgraphs depends on the average community length. It can also be observed that the size of the intermediate data is about 1–15 times of the size of the output data. The size of generated intermediate data plays a vital role in program efficiency. The larger the intermediate data, the longer the execution time. We also notice that for all the Supreme Court Citation datasets, much less intermediate data was generated compared to that for other datasets. For example, although the number of nodes and edges in the Supreme Court Citation dataset are around 9 times than those in the Bitcoin Alpha dataset, the generated intermediate data and the execution time of upper-level communities for the Supreme Court Citation are less than those of Bitcoin Alpha dataset. This is because our algorithm's complexity depends on the average degree of the nodes in the graph and the Bitcoin dataset has much higher average degree for the nodes. Furthermore, 99% and 87% of nodes in the Bitcoin Alpha dataset have developed five-level subgraphs in upper and lower parts, respectively; while 29,671 upper-level and 23,512 lower-level subgraphs are generated in the Supreme Court Citation dataset, which count for 86% and 68% of the total number of nodes in the graph, respectively. We can relate the above observations to the fact that in the

Supreme Court Citation dataset there exist more nodes with very low connectivity so that they cannot develop into larger communities, which results in smaller size of intermediate data and thus shorter execution time.

5 Conclusions

We have proposed a new parallel and scalable Map-Reduce based approach for discovering PageRank based hierarchical communities in large directed graphs. Our method is the first scalable Map-Reduce algorithm for community detection in directed graphs that constructs hierarchical structures around core nodes while maintaining edge directions. By utilizing the PageRank ranking algorithm, nodes with a similar rank as the core node are considered for inclusion in communities. We have successfully demonstrated that this method is especially suitable for finding communities in the graphs with the structure of the flow hierarchy, such as trust networks, citation networks and defeat networks.

References

1. Lai, D., Lu, H., Nardini, C.: Finding communities in directed networks by pagerank random walk induced network embedding. Physica A **389**(12), 2443–2454 (2010)
2. Satuluri, V., Parthasarathy, S.: Symmetrizations for clustering directed graphs. In: Proceedings of the 14th International Conference on Extending Database Technology, pp. 343–354. ACM (2011)
3. Gleich, D.: Hierarchical directed spectral graph partitioning. Inf. Netw. (2006)
4. Yen, L., Vanvyve, D., Wouters, F., Fouss, F., Verleysen, M., Saerens, M.: Clustering using a random walk based distance measure. In: ESANN, pp. 317–324 (2005)
5. Andersen, R., Chung, F., Lang, K.: Local graph partitioning using pagerank vectors. In: 2006 47th Annual IEEE Symposium on Foundations of Computer Science (FOCS 2006), pp. 475–486. IEEE (2006)
6. Rosvall, M., Bergstrom, C.T.: Maps of random walks on complex networks reveal community structure. Proc. Natl. Acad. Sci. **105**(4), 1118–1123 (2008)
7. Alamgir, M., Von Luxburg, U.: Multi-agent random walks for local clustering on graphs. In: 2010 IEEE International Conference on Data Mining, pp. 18–27. IEEE (2010)
8. Avrachenkov, K., Dobrynin, V., Nemirovsky, D., Pham, S.K., Smirnova, E.: Pagerank based clustering of hypertext document collections. In: Proceedings of the 31st Annual International ACM SIGIR Conference on Research and Development in Information Retrieval, pp. 873–874. ACM (2008)
9. Andersen, R., Chung, F., Lang, K.: Local partitioning for directed graphs using PageRank. In: Bonato, A., Chung, F.R.K. (eds.) WAW 2007. LNCS, vol. 4863, pp. 166–178. Springer, Heidelberg (2007). https://doi.org/10.1007/978-3-540-77004-6_13
10. Kumar, S., Spezzano, F., Subrahmanian, V., Faloutsos, C.: Edge weight prediction in weighted signed networks. In: 2016 IEEE 16th International Conference on Data Mining (ICDM), pp. 221–230. IEEE (2016)
11. Kumar, S., Hooi, B., Makhija, D., Kumar, M., Faloutsos, C., Subrahmanian, V.: Rev2: fraudulent user prediction in rating platforms. In: Proceedings of the Eleventh ACM International Conference on Web Search and Data Mining, pp. 333–341. ACM (2018)

Pattern Mining

FastTIRP: Efficient Discovery
of Time-Interval Related Patterns

Philippe Fournier-Viger[1]([⊠])[iD], Yuechun Li[1], M. Saqib Nawaz[1][iD],
and Yulin He[1,2][iD]

[1] Shenzhen University, Shenzhen, China
{philfv,msaqibnawaz}@szu.edu.cn, yulinhe@gml.ac.cn
[2] Guangdong Laboratory of Artificial Intelligence and Digital Economy (SZ),
Shenzhen, China

Abstract. Finding frequent patterns in discrete sequences of symbols or
events can be useful to understand data, support decision-making and
make predictions. However, many studies on analyzing event sequence
do not consider the duration of events, and thus the complex time relationships between them (e.g. an event may start at the same time as
another event but end before). To find frequent sequential patterns in
data where events have a start and end time, an emerging topic is time-
interval related pattern (TIRP) mining. Several algorithms have been
proposed for this task but efficiency remains a major issue due to the
very large search space. To provide a more efficient algorithm for TIRP
mining, this paper presents a novel algorithm called FastTIRP. It utilizes
a novel Pair Support Pruning (PSP) optimization to reduce the search
space. Experiments show that FastTIRP outperforms the state-of-the-art
VertTIRP algorithm in terms of runtime on four benchmark datasets.

Keywords: Time-intervals · Event sequences · Efficiency · Pair
support pruning · Discrete sequences

1 Introduction

A popular data representation is sequences. A sequence is a list of symbols or
events, and can encode various information such as a list of moves in a chess
game, and a list of events that occur in a computer network. To identify useful
knowledge that may be hidden in sequence data, several pattern mining algorithms have been designed. The aim is to detect patterns that appear frequently
in one or more sequences and satisfy some user-defined constraints. Some of
the most popular types of patterns are sequential patterns [1,6,14] and episodes
[3,5,9,10,15], which are respectively patterns found in a set of sequences, or in
a very long sequence.

Several algorithms have been proposed to find patterns in sequences, to help
users to understand the data and make prediction. However, most studies assume
a simple time representation, where each event is represented by a single timestamp. An example of sequence using this representation is shown in Fig. 1(a).

© The Author(s), under exclusive license to Springer Nature Switzerland AG 2022
P. P. Roy et al. (Eds.): BDA 2022 India, LNCS 13773, pp. 185–199, 2022.
https://doi.org/10.1007/978-3-031-24094-2_13

That sequence indicates that a person attended a meeting at 9:00 AM, made a phone call at 10:00 AM and then another phone call at 11:00 AM. A major problem with this time representation is that the duration of events is not taken into account (e.g. there is no information about the duration of the meeting or that of the phone calls). As a result, most algorithms can only find patterns involving two types of relationships between events: two events are either simultaneous or one appear before the other. For instance, a traditional sequential pattern mining algorithm could find a pattern indicating that a meeting is followed by a phone call.

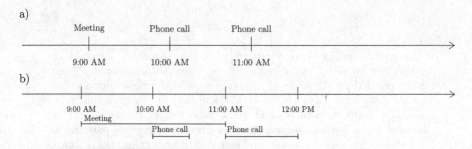

Fig. 1. A (a) simple event sequence and (b) a time-interval sequence

Recently, a richer time representation has been studied, called *time-interval sequence*, where events are represented as time-intervals with a start and end time. An example of time-interval sequence is shown in Fig. 1(b). It indicates that a person attended a meeting from 9:00 AM till 11:00 AM, and made a phone call *during* that meeting from 10:00 AM to 10:30 AM, and then made another phone call from 11:00 AM to 12:00 PM. It is easy to see that this representation shown in Fig. 1(b) carries more information than the one shown in Fig. 1(a). For instance, we can observe that the first phone call did not occur after the meeting (as it seemed in Fig. 1(a)), but during the meeting.

A popular task to find frequent patterns in such data is time-interval related pattern (TIRP) mining [8,11,12]. The goal is to find a form of sequential patterns called TIRP where events may be connected by various complex temporal relationships (i.e. an event overlaps with another event, an event starts at the same time as another event but end before, etc.). TIRP mining is a harder problem than traditional sequential pattern mining (SPM) [2] because each pair of events may be connected by multiple relationship types. TIRP mining has been used in several applications such as to analyze medical treatments [7], discover patterns in stock prices [13], and analyzing skating data [4].

Many algorithms have been proposed for TIRP mining such as ZMiner [8], VertTIRP [11], KarmaLego and DarmaLego [12]. However efficiency remains a major issue as the number of possible patterns is very large. To the best of our knowledge, the state-of-the-art algorithm is VertTIRP. It performs a depth-first search and relies on a vertical data structure to encode data about patterns.

The main drawback of VertTIRP is that it explores the search space by combining pairs of patterns and perform a costly join operation using vertical data structures to calculate their support.

In this paper, we aim to address the efficiency problem of TIRP mining by proposing a novel algorithm called FastTIRP. It is an enhanced version of VertTIRP where a novel technique called Pair Support Pruning (PSP) is applied. This technique consists of pre-calculating the frequency information about pairs of events to avoid performing join operations. Experiments on several benchmark datasets show that FastTIRP outperforms VertTIRP in terms of runtime as it performs less join operations, and also has good performance compared to KarmaLego.

The remaining sections of this paper are structured as follows. The problem definition of TIRP mining is explained in Sect. 2. The proposed FastTIRP algorithm is presented in Sect. 3. Section 4 provides the experiments results, followed by a conclusion in Sect. 5.

2   Problem Definition

The problem of discovering TIRPs is defined based on the following definitions, here presented using a similar notation as in the VertTIRP paper [11,12].

Definition 1 (Event type). *Let there be a set of event types $E = \{e_1, e_2, \ldots e_n\}$. Furthermore, assume that there exists a total order on these events denoted as $<$ which is the lexicographical order.*

Example 1. Consider a set of events $E = \{A, B, C\}$ containing three event types, and the lexicographical order defined as $A < B < C$.

Definition 2 (Symbolic time interval). *A symbolic time interval I is a triple of the form $I = (start, end, e)$ where start and end are numbers respectively indicating the start time and end time of an event of type $e \in E$. The notation $I.start$, $I.end$ and $I.e$ will be used in the following to refer to the elements **start**, **end** and **e** of a symbolic time interval I.*

To compare the timestamps of events, it is useful to use an approximate comparison of events so as to handle noise. This is done using two relationships, defined based on a user-defined epsilon ($\epsilon \geq 0$) value [12].

Definition 3 (Temporal relations between timestamps). *Let there be two timestamps t_i and t_j and a user-defined number ϵ. It is said that t_i and t_j are quasi-equal if $|t_i - t_j| \leq \epsilon$, which is denoted as $t_i =_\epsilon t_j$. Moreover, it is said that t_i precedes t_j if $|t_j - t_i| > \epsilon$, which is denoted as $t_i <_\epsilon t_j$ or $t_j >_\epsilon t_i$.*

Definition 4 (Time-interval sequence). *A time-interval sequence (TIS) S is an ordered list of symbolic time intervals $S = \langle I_1, I_2, \ldots I_q \rangle$. Symbolic time intervals within a TIS S are sorted according to a total order $\prec$ such that $I_i \prec I_j$ for any I_i and I_j if $(I_i.start <_\epsilon I_j.start) \vee (I_i.start =_\epsilon I_j.start \wedge I_i.end <_\epsilon I_j.end) \vee (I_i.start =_\epsilon I_j.start \wedge I_i.end =_\epsilon I_j.end \wedge I_i.e < I_j.e)$.*

Definition 5 (Time-interval sequence database). *A time-interval sequence database (TISD) D is a list of time-interval sequences* $D = \langle S_1, S_2, \ldots S_m \rangle$. *The sequence identifier of a sequence* $S_i \in D (1 \leq i \leq m)$ *is said to be i.*

Example 2. To illustrate these definitions, consider the TISD $D = \{S_1 : \langle(8, 12, A), (10, 12, B), (8, 11, C)\rangle$ $S_2 : \langle(8, 12, C), (10, 16, A), (15, 18, B)\rangle$ $S_3 : \langle(10, 16, C), (14, 19, B), (15, 19, A), (14, 16, B)\rangle\}$, which is represented visually in Fig. 2 [11]. This TISD contains three time-interval sequences called S_1, S_2 and S_3. The first sequence indicates that an event A started at time 8 and ended at time 12, an event B started at time 10 and ended at time 12, and an event C started at time 8 and ended at 11. The two other time-interval sequences can be interpreted in a similar way.

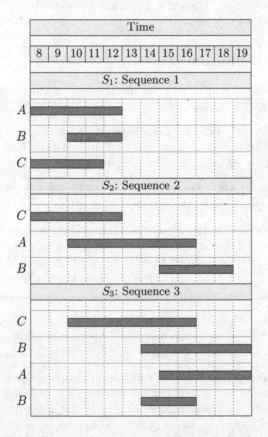

Fig. 2. An example time-interval sequence database

To be able to compare events, it is important to have a set of temporal relations. In this paper, the set of temporal relations from VertTIRP [11] is used. It is based on the 13 temporal relations introduced by Allen, with some modifications to handle epsilon [12] and resolve some ambiguities in the definitions used in prior work.

Definition 6 (Temporal relations between time intervals). *Let* $r(I_i, I_j)$ *be a function that determines the temporal relation between any pair of symbolic time intervals* I_i *and* I_j. *There are eight temporal relationships* $\omega = \{b, m, o, l,$ $c, f, e, s\}$, *called before, meets, overlaps, left contains, contains, is finished by, equal, and starts, respectively. Assume also that two optional constraints called* $mingap \geq 0$ *and* $maxgap \geq 0$ *are defined by the user. The function* $r(I_i, I_j)$ *returns*

$r(I_i, I_j) = \mathbf{b}$ *if* $(I_j.start - I_i.end) > \epsilon \wedge (I_j.start - I_i.end) < maxgap \wedge (I_j.start - I_i.end) > mingap$.

$r(I_i, I_j) = \mathbf{m}$ *if* $|I_j.start - I_i.end| \leq \epsilon \wedge (I_j.start - I_i.start > \epsilon \wedge (I_j.end - I_i.end) > \epsilon$

$r(I_i, I_j) = \mathbf{o}$ *if* $(I_j.start - I_i.start) > \epsilon \wedge (I_i.end - I_j.start > \epsilon \wedge (I_j.end - I_i.end) > \epsilon$

$r(I_i, I_j) = \mathbf{l}$ *if* $\epsilon > 0 \wedge |I_j.start - I_i.start| \leq \epsilon \wedge (I_i.end - I_j.end) > \epsilon$

$r(I_i, I_j) = \mathbf{c}$ *if* $(I_j.start - I_i.start) > \epsilon \wedge (I_i.end - I_j.end) > \epsilon$

$r(I_i, I_j) = \mathbf{f}$ *if* $(I_j.start - I_i.start) > \epsilon \wedge |I_j.end - I_i.end| \leq \epsilon$

$r(I_i, I_j) = \mathbf{e}$ *if* $|I_j.start - I_i.start| \leq \epsilon \wedge |I_j.end - I_i.end| \leq \epsilon$

$r(I_i, I_j) = \mathbf{s}$ *if* $|I_j.start - I_i.start| \leq \epsilon \wedge (I_j.end - I_i.end) > \epsilon$.

The problem of finding frequent time-interval related patterns (TIRPs) aims at finding TIRPs in a time-interval sequence database. A TIRP is defined as follows.

Definition 7 (Time-interval related pattern). *A time-interval related pattern (TIRP) X is a pair of the form* $X = (ev, rel)$ *where* **rel** *is an ordered list of temporal relations from* ω *and* **ev** *is an ordered list of events from E. Assume that* **ev** *contains* k *events. Then, the list* **rel** *indicates the temporal relations as follows: relations* $= \langle r(I_1, I_2), \ldots r(I_1, I_k), r(I_2, I_3), \ldots r(I_{k-1}, I_k) \rangle$. *In the following, the notation* $r_{i,j}$ *refers to* $r(e_i, e_j)$ *where* e_i *and* e_j *are the i-th and j-th events in* **ev**.

Example 3. Continuing the running example, the TIRP $(AB, \mathbf{o})$ occurs in sequence 2 and sequence 3 in the symbolic time intervals $\langle (10, 16, A), (15, 18, B) \rangle$ and $\langle (15, 19, A), (14, 16, B) \rangle$, respectively. The TIRP $(ABC, \mathbf{fso})$ appears in sequence 1 at $\langle (8, 12, A), (10, 12, B), (8, 11, C) \rangle$.

An important observation is that multiple TIRPs may have the same events but only differs by their temporal relations. This gives rise to the definition of Symbol-TIRP [11].

Definition 8 (Symbol-TIRP). *Let the notation* $\overline{ev}$ *denotes the set of all TIRPs having the same list of events ev.* $\overline{ev}$ *is said to be a Symbol-TIRP.*

Example 4. $\overline{BC} = \{(BC, \mathbf{o}), (BC, \mathbf{f})\}$ is the Symbol-TIRP containing the TIRPS having the list of events $ev = \langle BC \rangle$.

To find TIRPs in a database, it is necessary to rely on the concept of *match* between a TIRP and a time-interval sequence.

Definition 9 (TIRP matching). *Let there be a time-interval sequence* $S = \langle I_1, I_2, \ldots, I_q \rangle$ *and a TIRP* $X = (ev, rel)$. *Then,* X *is said to match* S *if* $\forall e_i, e_j \in ev$, $\exists I_\alpha, I_\beta \in S$ *such that* $e_i = I_\alpha.e$ *and* $e_j = I_\beta.e$ *and* $r(I_\alpha, I_\beta) = r_{i,j}$.

Example 5. Consider the time-interval sequence from the running example, $S = \langle (8, 12, A), (10, 12, B), (8, 11, C) \rangle$. For the TIRP $X = (ABC, \mathbf{fso})$, we have $I_1.e = A$, $I_2.e = B$, $I_3.e = C$, $r(I_1, I_2) = f$, $r(I_1, I_3) = s$ and $r(I_2, I_3) = o$. Hence, X is said to match S.

To find interesting TIRPs in a sequence, various interestingness function can be used. The two main functions are the horizontal support and vertical support [11].

Definition 10 (Horizontal support). *Let there be a TIRP* X *and a time-interval sequence* S. *The horizontal support of* X *in* S *is defined as* $h(X) = |S_X|$, *where* $|S_X|$ *is the total number of symbolic time intervals that match with* X.

Definition 11 (Vertical support). *Let there be a TIRP* X *and a time-interval sequence database* D. *The vertical support of* X *in* D *is defined as* $v(X) = |D_X|$, *where* $|D_X|$ *is the total number of sequences that match with* X.

Example 6. Continuing the running example, consider that $X = (AB, \mathbf{f})$. It can be found that X is matching with $\langle (8, 12, A) \rangle$ in sequence 1, and it also matches with $\langle (14, 19, B) \rangle$ in sequence 3. Thus, the vertical support of X is $v(X) = 2$.

Definition 12 (Problem of discovering all frequent TIRPS). *Let there be a time-interval sequence database* D, *a user-defined minimum support threshold* $minsup > 0$ *and some optional constraints:* $mingap \geq 0$, $maxgap \geq 0$, $\epsilon \geq 0$. *The problem of discovering all frequent TIRPS is to enumerate all frequent TIRPs, that is each TIRP* s *such that* $v(s) \geq minsup$ *under the constraints [11].*

Example 7. Back to the previous running example, if $minsup$ is set to 2, the set of frequent Symbol-TIRPs that are discovered is $R = \{(AB, 3, foo), (CA, 2, oo), (CB, 3, bof)\}$, where each TIRP is followed by its support and the temporal relations that in the order of their occurrence. For example, the TIRP AB with temporal relations f, o, o (finished by, overlaps, overlaps) has a vertical support of 3.

Note that it is possible to add some additional constraints such as the minimum and maximum duration of a TIRP (see [11]).

3   The FastTIRP Algorithm

This section presents the proposed FastTIRP algorithm. The motivation for developing this algorithm is that the state-of-art VertTIRP algorithm can have very long runtimes. After doing some preliminary analysis of the performance of VertTIRP, we found that its most costly operation is the join operation. This operation is used to calculate the support of a new pattern obtained by combining two existing patterns. To reduce the number of join operations, FastTIRP integrates a novel optimization called Pair Support Pruning (PSP) in VertTIRP. This optimization is based on a structure called Pattern Support Matrix (PSM).

The next subsections first explains briefly the search process and then describes the proposed PSP technique in details.

3.1 The Search Process

The pseudocode of the proposed FastTIRP algorithm is shown in Algorithm 1. The input is a time-interval sequence database D, a minimum support threshold $minsup$ and some optional parameters ϵ, $mingap$ and $maxgap$. The output is the set of all frequent TIRPs. The algorithm first configures the function $r(I_i, I_j)$ (to evaluate temporal relations between time-intervals) based on ϵ, $mingap$ and $maxgap$, and also configures the temporal relation $<_\epsilon$ (to evaluate temporal relations between timestamps) based on ϵ. Then, a function $FindFrequentEvents$ is called, which scans the input database to identify the set $FrequentEvents$ of all the TIRPs that are frequent and contain a single event. At the same time, the algorithm builds a vertical structure for each of these TIRPs. This structure is the same as in the VertTIRP algorithm [11] and is used to compute the support of each TIRP. After this, the algorithm scans the database to build the proposed PSM data structure, which will be presented in the next subsection. Then, the algorithm initializes a variable $Patterns$ to store all frequent TIRPs. Then a loop is done to try to extend each pattern p that is in $FrequentEvents$. This is done by calling a procedure called $Search$ with p, the list of frequent events $FrequentEvents$ that could be appended to p to form larger patterns, the minimum support threshold, and the PSM structure. The $Search$ function performs a depth-first search and returns all patterns that are frequent and obtained by appending events at the end of p. These patterns are then added to the set $Patterns$. Finally, the algorithm returns the set $Patterns$ containing all frequent TIRPs.

The search procedure is shown in Algorithm 2. The input is a Symbol-TIRP $\overline{p}$, a set $FrequentEvents$ containing Symbol-TIRPs each having a single event that could be appended to $\overline{p}$ to form larger TIRPs, the $minsup$ threshold, and the PSM structure. The procedure first initializes a set $NewPatterns$ and put the pattern $\overline{p}$ inside. Then, a variable $LocalFrequentEvents$ is initialized to the empty set, and will be used to store single events that are locally frequent. After that a loop is done on each frequent Symbol-TIRP $\overline{q}$ from the set $FrequentEvents$ to try to form a larger pattern by appending $\overline{q}$ at the end of $\overline{p}$ to create a larger Symbol-TIRP New. But before doing this step, the new pruning strategy PSP is checked, which will be presented in the next section. If that strategy determines that $\overline{p}$ and $\overline{q}$ cannot be combined, then this combination is skipped. Otherwise, the algorithm applies a join operation on the vertical structures of $\overline{p}$ and $\overline{q}$ to create the vertical structure of the resulting pattern New. If the support of the resulting pattern New is no less than $minsup$,

---

Algorithm 1: FastTIRP

---

input : D: a time-interval sequence database,
           minsup: the required minimum support,
           ϵ, *mingap*, and *maxgap*: optional parameters
output: all frequent TIRPs

1 Configure the temporal relations based on ϵ, *mingap* and *maxgap*;
2 $FrequentEvents = \texttt{FindFrequentEvents}(D, minsup)$;
3 $PSM = \texttt{BuildPSM}(D)$;
4 $Patterns = \emptyset$;
5 **foreach** $\overline{p} \in FrequentEvents$ **do**
6 $\quad$ $NewPatterns = \texttt{Search}(\overline{p}, FrequentEvents, minsup)$;
7 $\quad$ $Patterns = Patterns \cup NewPatterns$;
8 **end**
9 Return $Patterns$;

---

then it is added to the set of frequent TIRPs $NewPatterns$ and $\overline{q}$ is added to the set of locally frequent events $LocalFrequentEvents$. After that loop ends, another loop is done to recursively try to extend each frequent pattern $\overline{New}$ in $NewPatterns$. This is done by calling the $Search$ procedure with $\overline{New}$, the locally frequent events $LocalFrequentEvents$, $minsup$ and the PSM. Finally, all the patterns that have been found are returned by the search procedure.

The above description did not explain all the details of the algorithm such as how to create and join the vertical data structures of TIRPs. This is because these operations are the same as in the VertTIRP algorithm. Interested reader can refer to the paper describing VertTIRP for detailed explanations [11]. The difference between FastTIRP and VertTIRP is the addition of Line 3 in Algorithm 1 to build the PSM structure, and Line 5 in Algorithm 2 to use the PSM structure to avoid performing joins. The next subsection explains these differences in details.

3.2 The Pair Support Pruning Technique

The key difference between the proposed FastTIRP algorithm and VertTIRP is a novel data structure called Pattern Support Matrix (PSM) which is used to reduce the number of join operations that are performed. The PSM is built by reading the input database once and storing the co-occurrence frequency of each pair of events. Formally, the PSM can be defined as follows:

Definition 13 (Pair Support Matrix). *The Pair Support Matrix of a time-interval sequence database D is denoted as PSM. For each pair of event types $e_1, e_2 \in E$, the PSM stores a triple of the form $(e_1, e_2, h(e_1, e_2))$ where $h(e_1, e_2)$ is the number of symbolic time intervals containing $e_1 <_\epsilon e_2$. In case the user utilizes the maximum gap constraint maxgap, then $h(e_1, e_2)$ is defined as the number of symbolic time intervals containing $e_1 <_\epsilon e_2$ within a time maxgap.*

Algorithm 2: Search

---

input : $\overline{p}$: a Symbol-TIRP,
            FrequentEvents: a set of Symbol-TIRPs having a single event,
            minsup: the minimum support threshold
output: $\overline{p}$ and all frequent TIRPs that extend $\overline{p}$

1   $NewPatterns = \{\overline{p}\}$;
2   $LocalFrequentEvents = \emptyset$;
3   **foreach** $\overline{q} \in FrequentEvents$ **do**
4   // Check the PSP pruning condition
5   **if** CheckPSP$(\overline{p}, \overline{q}, minsup)$ **then**
6          $New = $ Join$(\overline{p}, \overline{q})$;
7   **if** $sup(New) \geq minsup$ **then**
8             $NewPatterns = NewPatterns \cup \{New\}$;
9             $LocalFrequentEvents = LocalFrequentEvents \cup \{\overline{q}\}$;
10   **end**
11   **end**
12 **end**
13 **foreach** $\overline{New} \in NewPatterns$ **do**
14      $Extensions = $ Search$(\overline{New}, LocalFrequentEvents, minsup, PSM)$;
15      $NewPatterns = NewPatterns \cup Extensions$;
16 **end**
17 **return** $NewPatterns$;

There are several ways of implementing the PSM structure. The most simple way is to define it as a two dimensional matrix where there is a row and column for each event. Then, for a given column and row representing some events e_1 and e_2; the content of the cell is the number of symbolic time intervals containing $e_1 <_\epsilon e_2$. As example, Table 1 shows the PSM structure built for the running example implemented as a two dimensional matrix. The PSM structure contains several values indicating for instance that event type B appears in one time-interval before A, but A appears in 3 symbolic time-intervals before B.

Table 1. The PSM structure as a full matrix

	A	B	C
A	0	1	2
B	3	1	3
C	1	1	0

To reduce memory, the PSM can be also implemented as a triangular matrix as shown in Table 2. In this case, for any two events $e_1, e_2 \in E$, the two cells $(e_1, e_2, h(e_1, e_2))$ and $(e_2, e_1, h(e_2, e_1))$ are merged as: $(e_1, e_2, h(e_1, e_2) + h(e_2, e_1))$. This is the implementation used in the experiments of this paper.

But note that the full matrix or triangular matrix can still contain many zeros. In this case, the PSM could also be implemented as a sparse matrix (e.g. using a hash map of hash maps) to avoid storing cells containing zeros. This could be beneficial for datasets with a very large number of event types. It is also possible to redefine the PSM to store the vertical support instead of the horizontal support.

Table 2. The PSM structure as a triangular matrix

	A	B	C
A	0	4	3
B		1	4
C			0

The PSM structure is used to reduce the number of join operations done by FastTIRP as follows. Before joining an event $\overline{q}$ with a Symbol-TIRP $\overline{p}$ in Line 6 of Algorithm 2, FastTIRP checks the support of the last event of $\overline{p}$ with event $\overline{q}$ in the PSM. If that value is no greater than $minsup$, it is unnecessary to append $\overline{q}$ to p to form a larger pattern, as the result will be an infrequent TIRP. Thus, the join operation is not performed.

For example, consider that $minsup = 4$, $\overline{p} = \overline{BC}$, which matches with $\langle (10, 16, C), (14, 19, B) \rangle$ and that $\overline{q} = \overline{B}$, which matches with $C(15, 18, B)$. By looking at the PSM, we can find that the support of B with B is 1 and thus that $\overline{p}$ should not be combined with $\overline{q}$ to obtain a Symbol-TIRP $\overline{CBB}$.

The PSM technique can be used to effectively reduce the number of join operations as it will be demonstrated in the experiments. The design of PSP is inspired by a similar structure called CMAP used in sequential pattern mining [1].

4   Experimental Evaluation

This section presents the experimental evaluation of the proposed FastTIRP algorithm and obtained results are discussed. In experiments, the FastTIRP was compared with two algorithms for TIRP mining, which are: VertTIRP [11] and KarmaLego [12]. A prior study [11] compared VertTIRP with KarmaLego and obtained results show that VertTIRP was on overall faster than KarmaLego. However, in the experiment, the two algorithms were implemented in different programming languages (Python and C#, respectively), which may make the results unreliable. To avoid this issue, in this work, the three algorithms are implemented in C#. The C# implementations of VertTIRP and KarmaLego were obtained from: http://github.com/TIRPClo. The FastTIRP algorithm was implemented by modifying the C# code of VertTIRP.

All experiments were run on a laptop running Windows 11, with 16 GB of RAM and an 11th generation Core i7-11800H processor. Four datasets, namely *Diabetes*, *Hepatitis*, *Smarthome* and *ASL*, having various characteristics were used to evaluate the performance of the three algorithms. The first two datasets contain medical data, while the third and fourth datasets contain smarthome events and sign language utterances transcribed from video recordings, respectively. The main characteristics of the four datasets are listed in Table 3. Diabetes (Hepatitis) and Smarthome (ASL) contain 2,038 (498) and 89 (65) entities (sequences), 80,538 (48,029) and 23,213 (2,037) symbolic time intervals, and 35 (63) and 95 (146) event (symbol) types respectively. These datasets were downloaded from http://github.com/TIRPClo. For all experiments, epsilon was set to zero, *mingap* to 0 and *maxgap* to 30.

Table 3. Characteristics of the datasets

Dataset	Sequences	Time intervals	Event types
Diabetes	2,038	80,538	35
Hepatitis	498	48,029	63
Smarthome	89	23,213	95
ASL	65	2,037	146

4.1 Influence of *minsup* on Runtime, Number of Joins and Patterns

In this section, experiments were carried out to first evaluate the efficiency of the algorithms in terms of runtime and number of joins operations that are performed. Both runtime and number of joins were measured while varying the *minsup* parameter for each dataset. Figure 3 shows the execution time (left side) of the three algorithms and the number of joins operations (right side) for two algorithms (FastTIRP, VertTIRP) on four datasets.

For runtime, it is observed that FastTIRP was faster, overall, than VertTIRP on all datasets. For lower *minsup* values, the difference in runtime is higher. However, the runtime difference between FastTIRP and VertTIRP tend to decrease as *minsup* is increased. Interestingly, on all datasets, FastTIRP was about 20% faster than VertTIRP. On the other hand, KarmaLego performed better than FastTIRP on two datasets: Hepatitis and ASL. KarmaLego was indeed faster than FastTIRP for lower *minsup* values. For the Diabetes and ASL datasets, the average execution times for KarmaLego was high. That is why they are not added in the figure. KarmaLego took more than half hour (the time limit we set for runtime execution).

The number of joins performed by FastTIRP and VertTIRP is also observed to assess the ability of the PSM technique at reducing the number of joins. FastTIRP performed less number of joins than VertTIRP on all datasets (Fig. 3). This shows that the PSM technique in FastTIRP was able to reduce the number of joins in the Diabetes, Hepatitis, Smarthome and ASL datasets by up to 24 %, 61 %, 94%, and 91%, respectively.

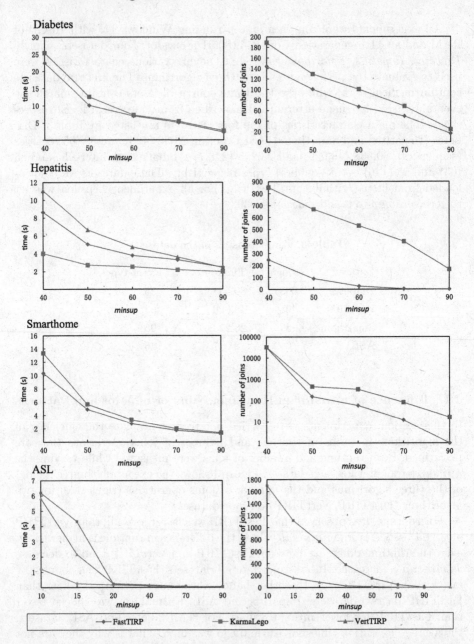

Fig. 3. Comparison of runtime and number of joins for different *minsup* values on four detests

Next, we investigate the number of TIRPs discovered by the FastTIRP algorithm on four datasets for the smallest and largest *minsup* values used in the experiments. The results are listed in Table 4. More TIRPs were found in the

Diabetes dataset, followed by Hepatitis, Smarthome and ASL. The main reason for this is that Diabetes contain more sequences and time intervals, followed by Hepatitis, Smarthome and ASL respectively. Moreover, more (less) TIRPs can be discovered by decreasing (increasing) the *minsup* value.

Table 4. Number of frequent TIRPs found in each dataset

Dataset	*minsup* range	Number of TIRPs
Diabetes	[40, 90]	[69692, 30094]
Hepatitis	[40, 90]	[44148, 25802]
Smarthome	[40, 90]	[19944, 4684]
ASL	[10, 90]	[1807, 72]

4.2   Influence of *minsup* on the Overall Memory Usage

This section analyzes the memory used by algorithms during execution. The results for the three algorithms are listed in Table 5.

FastTIRP used less memory compared to VertTIRP on the Diabetes dataset that contain more sequences and time intervals. On the Smarthome and ASL datasets, that contain less sequences, the memory usage of FastTIRP and VertTIRP is almost similar. Interestingly, the memory usage of FastTIRP and VertTIRP on the Smarthome dataset was higher than on the Hepatisis dataset. However, the Smarthome dataset has less sequences and time intervals than the Hepatitis dataset. KarmaLego performed better than FastTIRP and VertTIRP on the Diabetes and ASL datasets. KarmaLego's results for Hepatitis and Smarthome are not added in the Table 5 as it was unable to terminate for various *minsup* values within the time limit of half hour that we set for termination.

Table 5. Comparison of the memory usage on the four datasets

Dataset	Algorithm	Avg. memory usage (MB)	Max. memory usage (MB)
Diabetes	FastTIRP	559	687
	VertTIRP	538	663
	KarmaLego	280	284
Hepatitis	FastTIRP	283	368
	VertTIRP	280	364
Smarthome	FastTIRP	437	634
	VertTIRP	437	630
ASL	FastTIRP	213	216
	VertTIRP	212	215
	KarmaLego	29	35

We can further analyze the memory usage of FastTIRP by calculating the size of the proposed PSM data structure. Let $|E|$ be the number of event types. If the PSM is implemented as a full matrix, then the size can be calculated as $SizeFullMatrix = |E|^2 \times SizeOfValue$, that is the number of events to the power of two, multiplied by the number of bytes required to store a support value. In the C# implementation of FastTIRP, support values are stored using integers of four bytes. Thus, the size of the PSM as a full matrix for each dataset is as follows: Diabetes: $35^2 \times 4$ bytes= $4,900$ bytes, Hepatitis: $63^2 \times 4$ bytes= $15,876$ bytes, Smarthome: $95^2 \times 4$ bytes= $36,100$ bytes, and ASL: $146^2 \times 4$ bytes= $85,264$ bytes. In all cases, the size is quite small (less than a megabyte), which is acceptable.

For datasets with a very large number of event types, it may be worthwhile to implement the PSM as a triangular matrix or sparse matrix to save some memory. As a triangular matrix, the size of the PSM is $SizeTriangularMatrix = \frac{|E| \times (|E|-1)}{2} \times SizeOfValue$, and the PSM requires the following amount of memory for each dataset: Diabetes: $\frac{35 \times 34}{2} \times 4$ bytes= 2380 bytes, Hepatitis: $\frac{63 \times 62}{2} \times 4$ bytes= $7,812$ bytes, Smarthome: $\frac{95 \times 94}{2} \times 4$ bytes= $17,860$ bytes, and ASL: $\frac{146 \times 145}{2} \times 4$ bytes= $42,340$ bytes. This is the implementation used in experiments.

5   Conclusion

This paper has presented a novel algorithm for mining TIRPs in a time-interval sequence database, called FastTIRP. It utilizes the same basic search procedure as VertTIRP but applies a Pair Support Pruning (PSP) technique to reduce the number of join operations. Experiments show that FastTIRP outperforms the state-of-the-art VertTIRP algorithm in terms of runtime on several benchmark datasets and that the number of join operations can be greatly reduced.

In future work, we will consider designing other optimizations and algorithms for this problem, as well as a distributed version of FastTIRP to run on a big data framework.

References

1. Fournier-Viger, P., Gomariz, A., Campos, M., Thomas, R.: Fast vertical mining of sequential patterns using co-occurrence information. In: Tseng, V.S., Ho, T.B., Zhou, Z.-H., Chen, A.L.P., Kao, H.-Y. (eds.) PAKDD 2014. LNCS (LNAI), vol. 8443, pp. 40–52. Springer, Cham (2014). https://doi.org/10.1007/978-3-319-06608-0_4
2. Fournier-Viger, P., Lin, J.C.W., Kiran, U.R., Koh, Y.S.: A survey of sequential pattern mining. Data Sci. Pattern Recognit. **1**(1), 54–77 (2017)
3. Fournier-Viger, P., Yang, Y., Yang, P., Lin, J.C.-W., Yun, U.: TKE: mining top-K frequent episodes. In: Fujita, H., Fournier-Viger, P., Ali, M., Sasaki, J. (eds.) IEA/AIE 2020. LNCS (LNAI), vol. 12144, pp. 832–845. Springer, Cham (2020). https://doi.org/10.1007/978-3-030-55789-8_71

4. Huang, J.W., Jaysawal, B.P., Chen, K.Y., Wu, Y.B.: Mining frequent and top-k high utility time interval-based events with duration patterns. Knowl. Inf. Syst. **61**(3), 1331–1359 (2022)
5. Iwanuma, K., Takano, Y., Nabeshima, H.: On anti-monotone frequency measures for extracting sequential patterns from a single very-long data sequence. In: Conference on Cybernetics and Intelligent Systems, vol. 1, pp. 213–217 (2004)
.6. Jiang, C., Coenen, F., Zito, M.: A survey of frequent subgraph mining algorithms. Knowl. Eng. Rev. **28**, 75–105 (2013)
7. Le, H.H., Kushima, M., Araki, K., Yokota, H.: Differentially private sequential pattern mining considering time interval for electronic medical record systems. In: International Database Applications and Engineering Symposium, pp. 13:1–13:9 (2019)
8. Lee, Z., Lindgren, T., Papapetrou, P.: Z-miner: an efficient method for mining frequent arrangements of event intervals. In: SIGKDD Conference on Knowledge Discovery and Data Mining, pp. 524–534. ACM (2020)
9. Liao, G., Yang, X., Xie, S., Yu, P.S., Wan, C.: Mining weighted frequent closed episodes over multiple sequences. Tehnički vjesnik **25**(2), 510–518 (2018)
10. Mannila, H., Toivonen, H., Verkamo, A.I.: Discovering frequent episodes in sequences. In: International Conference on Knowledge Discovery and Data Mining, pp. 210–215. AAAI Press (1995)
11. Mordvanyuk, N., López, B., Bifet, A.: vertTIRP: robust and efficient vertical frequent time interval-related pattern mining. Expert Syst. Appl. **168**, 114276 (2021)
12. Moskovitch, R., Shahar, Y.: Classification of multivariate time series via temporal abstraction and time intervals mining. Knowl. Inf. Syst. **45**(1), 35–74 (2015)
13. Wu, S.Y., Chen, Y.L.: Mining nonambiguous temporal patterns for interval-based events. IEEE Trans. Knowl. Data Eng. **19**(6), 742–758 (2007)
14. Zheng, Z., Wei, W., Liu, C., Cao, W., Cao, L., Bhatia, M.: An effective contrast sequential pattern mining approach to taxpayer behavior analysis. World Wide Web **19**(4), 633–651 (2016)
15. Zhou, W., Liu, H., Cheng, H.: Mining closed episodes from event sequences efficiently. In: Zaki, M.J., Yu, J.X., Ravindran, B., Pudi, V. (eds.) PAKDD 2010. LNCS (LNAI), vol. 6118, pp. 310–318. Springer, Heidelberg (2010). https://doi.org/10.1007/978-3-642-13657-3_34

Discovering Top-k Periodic-Frequent Patterns in Very Large Temporal Databases

Palla Likhitha[✉][iD], Penugonda Ravikumar[iD], Rage Uday Kiran[iD],
and Yutaka Watanobe[iD]

University of Aizu, Aizuwakamatsu, Japan
likhithapalla7@gmail.com, raviua138@gmail.com,
{udayrage,yutaka}@u-aizu.ac.jp

Abstract. Discovering periodic-frequent patterns in temporal databases is a challenging data mining problem with abundant applications. It involves discovering all patterns in a database that satisfy the user-specified *minimum support* (*minSup*) and *maximum periodicity* (*maxPer*) constraints. *MinSup* controls the minimum number of transactions in which a pattern must appear in a database. *MaxPer* controls the maximum time interval within which a pattern must reappear in the database. Setting an appropriate *minSup* and *maxPer* values for any given database is an open research problem. This paper addresses this open problem by proposing a solution to discover top-k periodic-frequent patterns in a temporal database. Top-k periodic-frequent patterns represent a total of k periodic-frequent patterns with the lowest *periodicity* value in a database. An efficient depth-first search algorithm, called Top-k Periodic-Frequent Pattern Miner (k-PFPMiner), which takes only k threshold as an input was presented to find all desired patterns in a database. Experimental results on synthetic and real-world databases demonstrate that our algorithm is memory and runtime efficient and highly scalable.

Keywords: Data mining · Pattern mining · Temporal databases

1 Introduction

Frequent pattern mining [1] is a popular data mining model aiming to discover all frequently occurring patterns in a transactional database. However, a fundamental limitation of this model is that it fails to discover temporal regularities that may exist in a temporal database. When confronted with this problem in real-world applications, researchers proposed an extended model to discover periodic-frequent patterns [12] in a temporal database that satisfy the user-specified *minimum support* (*minSup*) and *maximum periodicity* (*maxPer*) constraints.

Several pattern mining techniques, such as fuzzy periodic-frequent pattern mining [6], local periodic pattern mining [4], partial periodic pattern mining

© The Author(s), under exclusive license to Springer Nature Switzerland AG 2022
P. P. Roy et al. (Eds.): BDA 2022 India, LNCS 13773, pp. 200–210, 2022.
https://doi.org/10.1007/978-3-031-24094-2_14

[9], stable periodic pattern mining [5], recurring pattern mining [8], and periodic sequential pattern mining, were inspired from the periodic-frequent pattern model. However, this model's widespread adoption and successful industrial application was hindered by this obstacle: "*minSup and maxPer are two key constraints that make periodic-frequent pattern mining practicable in real-world applications. They are used to prune the search space and limit the number of patterns generated. Unfortunately, setting these two constraints for an application is an open research problem and may require a profound knowledge of the application background.*" This paper addresses this challenging open problem by finding top-k periodically occurring frequent patterns in a temporal database. This paper's contribution is as follows: (i) we proposed a novel model of top-k periodic-frequent patterns that may exist in temporal databases. Only one constraint k is used to find the interesting top-k periodic-frequent patterns with the lowest *periodicity* in the entire database. (ii) We also introduced an upper-bound measure and a pruning technique called *dynamic maximum periodicity* to reduce the search space and in pruning the uninteresting patterns. (iii) We proposed an efficient search algorithm called top-k Periodic-frequent Pattern Miner (k-PFPMiner) to find all the desired patterns. (v) Experimental results on synthetic and real-world databases demonstrate that our algorithm is memory and runtime efficient and highly scalable.

The rest of the paper is organized as follows. Section 2 describes related work on finding top-k periodic-frequent patterns in databases. Section 3 describes a proposed model to find top-k periodic-frequent patterns in databases. Section 4 describes the proposed algorithm to discover the top-k periodic-frequent patterns. Section 5 presents the experimental results obtained. Finally, in Sect. 6, we conclude and discuss future research.

2 Related Work

Agrawal et al. [1] introduced the concept of frequent pattern mining to extract useful information from transactional databases. Luna et al. [10] conducted a detailed survey on frequent pattern mining and presented the improvements in the past 25 years. However, finding patterns that appear regularly in a database with the help of frequent pattern mining, which only considers frequency, is not appropriate.

Tanbeer et al. [12] generalized the frequent pattern model to discover periodic-frequent patterns in a temporal database. Amphawan et al. [2] introduced a model to find the most periodic-frequent patterns called Top-k periodic-frequent patterns. Uday et al. [7] have designed a novel concept named *local periodicity* to prune the non-periodic patterns locally. Authors have discarded the patterns whose *local periodicity* is less than the user-specified *maxPer* value. As a result, most of the non-periodic patterns tid-lists were not completely built, resulting in a decrease in the computational time of the proposed algorithm. Since its inception, the problem of finding periodic-frequent patterns has received a great deal of attention [3,11]. The basic model used in most of these algorithms remains the same. Discovering complete set of patterns in

Table 1. Temporal database

ts	items	ts	items
1	pqr	6	prs
2	qrs	7	pqrst
3	pqrs	8	pr
4	pqrt	9	pqr
5	qr	10	st

temporal databases that satisfy the user-specified $minSup$ and $maxPer$ values. Setting a user-specified $minSup$ and $maxPer$ for a given database is an open research problem. When confronted with this problem in real-world applications, researchers have tried to find top-k periodic-frequent patterns. Amphawan et al. [2] described the top-k periodic-frequent patterns from transactional databases without $minSup$ constraint. However, the authors have used the user-specified $maxPer$ constraint and a k value to generate top-k periodic-frequent patterns.

In this paper, we are updating the $maxPer$ value dynamically without user intervention while generating exciting patterns. By setting the user-defined k value only, we are extracting the top-k periodic-frequent patterns from the large temporal databases.

3   Proposed Model: top-k Periodic-Frequent Patterns

Let O be the set of objects (or items). Let $P \subseteq O$ be a **itemset** (or a pattern). A itemset containing α, $\alpha \geq 1$, number of items is called a α-**pattern**. In a **transaction**, $t_k = (ts, X)$ is a tuple, where ts represents the timestamp at which the pattern X has occurred. A **temporal database** TDB over O is a set of transactions, i.e., $TDB = \{tr_1, \cdots, tr_m\}$, $m = |TDB|$, where $|TDB|$ can be defined as the number of transactions in TDB. For a transaction $tr_k = (ts, X)$, $k \geq 1$, such that $Z \subseteq X$, it is said that Z occurs in tr_k (or tr_k contains Z) and such a timestamp is denoted as ts^Z. Let $TS^Z = \{ts_j^Z, \cdots, ts_k^Z\}$, j, $k \in [1, m]$ and $j \leq k$, be an **ordered set of timestamps** where Z has occurred in TDB.

Example 1. Let $O = \{p, q, r, s, t\}$ be the set of items. A temporal database generated from items O is shown in Table 1. The set of items 'r' and 'p', i.e., $\{r, p\}$ is considered as a pattern. For short, we represent this pattern as 'rp.' This pattern is a 2-pattern because it contains two items. The pattern 'rp' appears at the timestamps of 1, 3, 4, 6, 7, 8, and 9. Therefore, the list of timestamps containing 'rp', i.e., $TS^{rp} = \{1, 3, 4, 6, 7, 8, 9\}$.

Definition 1. *(**Periodicity of** Z) A period of Z in TDB is calculated using the following three ways: (i) $p_1^Z = ts_a^Z - ts_{min}$, (ii) $p_i^Z = ts_q^Z - ts_p^Z$, where $2 \leq i \leq |TS^Z|$ and $a \leq p \leq q \leq c$ represent the periods (or inter-arrivals) of Z in the database, and (iii) $p_{|TS^Z|+1}^Z = ts_{max} - ts_c^Z$. The maximal and minimal timestamps of all transactions in the database are represented as ts_{min} and*

ts_{max}. Let $P^Z = \{p_1^Z, p_2^Z, \cdots, p_k^Z\}, k = |TS^Z| + 1$, be the set of all periods of Z in $UTDB$. The periodicity of Z, denoted as $per(Z) = max(p_1^Z, p_2^Z, \cdots, p_k^Z)$.

Example 2. The periods for this pattern are: $p_1^{rp} = 1\ (= 1 - ts_{initial})$, $p_2^{rp} = 2\ (= 3 - 1)$, $p_3^{rp} = 1\ (= 4 - 3)$, $p_4^{rp} = 2\ (= 6 - 4)$, $p_5^{rp} = 1\ (= 7 - 6)$, $p_6^{rp} = 1\ (= 8 - 7)$, $p_7^{rp} = 1\ (= 9 - 8)$, and $p_8^{rp} = 1\ (= ts_{final} - 9)$, where $ts_{initial} = 0$ represents the timestamp of initial transaction and $ts_{final} = |TDB| = 10$ represents the timestamp of final transaction in the database. The *periodicity* of rp, i.e., $per(rp) = maximum(1, 2, 1, 2, 1, 1, 1, 1) = 2$.

Definition 2. *(Top-K periodic-frequent pattern X.)* Let $\{X_1, X_2, \cdots, X_k, \cdots, X_p\}$, $1 \le k \le p \le 2^m - 1$, be an ordered set of all patterns such that $per(X_1) \le per(X_2) \le \cdots \le per(X_k) \le \cdots \le per(X_p)$. A pattern X_a, $1 \le a \le p$, is said to be a top-k periodic pattern if its periodicity is no more than the periodicity of pattern X_k in the database. That is, X_a is said to be a top-k periodic pattern if $per(X_a) \le per(X_k)$.

Example 3. If the $per(rp) \le per(t)$, the pattern rp is considered as top-k periodic frequent patterns.

Definition 3. *(Problem definition.)* Given a temporal database (*TDB*) and a user-specified k value, the goal of top-k periodic pattern mining is to discover only top-k periodic-frequent patterns that have the lowest periodicities in a database.

4 Our Algorithm

4.1 Basic Idea: Dynamic Maximum Periodicity

Reducing the enormous search space is challenging as our model does not employ any constraint to reduce the search space. In this context, our idea to reduce this huge search space is as follows: "*Create an empty list known as* candidate periodic pattern-list *(or cpp-List). Also, create a Max-heap data structure and set its root to null. Then, scan the database and keep adding the patterns to the cpp-List. Simultaneously, update the Max-heap with the periodicity values of those items. Once the size of cpp-List reaches the size of k, set* dynamic maximum periodicity *(dMaxPer) equal to the value of root in Max-heap. Prune the search space (or itemsets) using the dMaxPer constraints. If we find any pattern in the constraint, add the corresponding pattern into the cpp-List by removing the existing k-pattern. We will update dMaxPer accordingly. We keep repeating this process until we complete the search space.*" The time complexity to determine *periodicity* of a pattern are $O(1)$ and $O(n)$, respectively. Where n represents the number of timestamps (or *frequency*) of a pattern in the database.

Definition 4. *(Dynamic maximum periodicity constraint.)* Let $AP = \{X_1, X_2, \cdots, X_{2^n-1}\}$, $n \ge 1$, be the set of all patterns in a database. Let $EP \subseteq SP$ be the set of patterns explored by our algorithm until now. Let $EP_k \subseteq EP$ such that $|EP_k| = k$ be a set of top-k candidate periodic patterns *found until now. The* dynamic maximum periodicity, *denoted as dMaxPer, represents the highest periodicity among all patterns in* EP_k. *That is, dMaxPer =* $\{max(per(X_p)|\forall X_p \in EP_k)\}$.

Example 4. Let $AP = \{p, q, r, s, t, pq, pr, \cdots, pqrst\}$ be the set of all patterns in a database. Let $EP = \{p, q, r, s, t\} \subset AP$ be the set of patterns explored until now. If $k = 5$, then $EP_k = \{p, q, r, s, t\}$. Thus, $dMaxPer = max(per(p), per(q), per(r), per(s), per(t)) = max(2, 2, 1, 3, 4) = 4$. For the pattern pr, $TS^{pr} = \{1, 3, 4, 6, 7, 8, 9\}$ and $per(pr) = 2$. If we explore a different pattern pr, then $EP = \{p, q, r, s, t, pr\}$. As $per(pr) < dMaxPer$, we prune pattern t and add pr in the EP_k. Thus, $EP_k = \{p, q, r, s, pr\}$ and $dMaxPer = max(2, 2, 1, 3, 2) = 3$. Thus, $dMaxPer$ automatically gets updated whenever a candidate top-k periodic-frequent pattern is found in the database.

The above constraint says that a pattern must have *periodicity* less than the $dMaxPer$ to be a candidate top-k periodic pattern. Thus, this constraint can be used to determine the minimal occurrences a pattern must have to be a candidate top-k periodic-frequent pattern.

Property 1. (**Pruning technique:**) Prune the pattern X if $per(X) > dMaxPer$. It is because neither X nor its supersets can be top-k periodic-frequent patterns.

The correctness of this property is based on Properties 2, 3, and Lemma 1. Our algorithm uses the above pruning technique to discover top-k periodic patterns effectively.

Property 2. For a pattern X, if $per(X) > dMaxPer$, then X cannot be a top-k periodic pattern.

Lemma 1. *For a pattern X, if $per(X) > dMaxPer$, then X cannot be a top-k periodic pattern.*

Proof. The correctness is straight forward to prove from Property 2.

Property 3. If $X \subset Y$, then $per(X) \leq per(Y)$ as $TS^X \supseteq TS^Y$.

4.2   *k*-PFPMiner

k-PFPMiner starts with finding top-k single items in the database and stores them in cPP-List described in Algorithm 1. Next, Algorithm 2 describes the procedure for finding top-k periodic patterns in a depth-first search manner. We now describe the working of this algorithm using the newly generated cPP-list.

   We start with item r, the pattern in the cPP-list with the lowest periodicity (line 2 in Algorithm 2). We preserve the *periodicity* of r, as shown in Fig. 1(a). Since r is a periodic-frequent pattern, we proceed to its child node rp by combining with other periodic items in the database and generate its TS-list by intersecting TS-lists of both items r and p, i.e., $TS^{rp} = TS^r \cap TS^p$ (lines 3 and

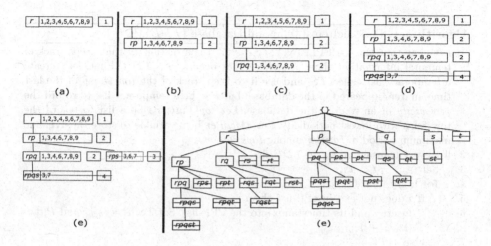

Fig. 1. Mining Top-K periodic patterns using DFS

4 in Algorithm 2). We record *periodicity* of rp, as shown in Fig. 1(b). We verify whether rp is a candidate periodic-frequent pattern or uninteresting pattern (line 6 in Algorithm 2). Since rp is a candidate periodic-frequent pattern, we check if *periodicity* of rp if $per(pr) < dMaxPer$, (in Algorithm 3) and calculate $dMaxPer$ as a maximum of all the *periodicities* of periodic-frequent patterns in current existing *topkPatterns*. Since rp is a top-k periodic-frequent pattern, we proceed to its child node rpq and generate its TS-list by performing the intersection of TS-lists of rp and q, i.e., $TS^{rpq} = TS^{rp} \cap TS^q$. We record *periodicity* of rpq, as shown in Fig. 1(c), identify it as a periodic-frequent pattern and check if it can be a top-k periodic frequent pattern. Since rpq is a top-k periodic frequent pattern, we again move to its child node $rpqs$ and generate its TS-list by performing the intersection of TS-lists of rpq and s, i.e., $TS^{rpqs} = TS^{rpq} \cap TS^s$. As *periodicity* of $rpqs$ is greater than the $dMaxPer$, we will prune the pattern $rpqs$ from the candidate periodic patterns list as shown in Fig. 1(d). We move to the other child of rp and generate its TS-list by performing the intersection of TS-lists rp and s, i.e., $TS^{rps} = TS^{rp} \cap TS^s$. As the *periodicity* of rps is greater than the $dMaxPer$, we will prune the pattern rps from the candidate periodic-frequent patterns list as shown in Fig. 1(e). The exact process is done for all the remaining nodes in the tree to find all periodic-frequent patterns. The complete set of periodic-frequent patterns generated from Table 1 is shown in Fig. 1(f) without striking. The above approach of finding periodic-frequent patterns using the downward closure property is efficient because it effectively reduces the search space and the computational cost.

5 Experimental Results

Since there exists no algorithm to find Top-k periodic-frequent patterns in temporal databases with only k constraint, we evaluated our algorithm k-PFPMiner on different databases varying k.

Algorithm 1. PeriodicItems(Temporal Database (TDB), K (k):

1: Let's say that the T-PFPList=$(Y, TS\text{-}list(Y))$ is a dictionary that keeps track of temporal information about a pattern that occurs in a TDB. First, let's create a temporary list called TS_l and use it to keep track of the *timestamp* of the last time an item appeared in the database. Let Per be a temporary list to record the *periodicity* of an item in the database. Let $topkPatterns$ be a list to record the top items with lowest periodicity. Let $dMaxPer$ be a variable to store the dynamic maximum period $dMaxPer$ among $topkPatterns$.

2: **for** every transaction $t_{cur} \in TDB$ **do**

3:     Set $ts_{cur} = t_{cur}.ts$;

4:     **for** every item $i \in t_{cur}.X$ **do**

5:         **if** i does not exit in cPP-list **then**

6:             Insert i and its timestamp into the PFP-list. Set $TS_l[i] = ts_{cur}$ and $P[i] = (ts_{cur} - ts_{initial})$;

7:         **else**

8:             Add i's timestamp in the cPP-list. Update $TS_l[i] = ts_{cur}$ and $P[i] = max(P[i], (ts_{cur} - TS_l[i]))$;

9: **for** each item i in cPP-list **do**

10:     Calculate $P[i] = max(Per[i], (ts_{final} - TS_l[i]))$;

11: Sort the remaining items in the cPP-list in ascending order of their *periodicity*.

12: **for** each item i in PFP-list **do**

13:     **if** $length(topkPatterns) < K$: **then**

14:         Store the item into $topkPatterns$

15: $dMaxPer = max$(periodicity of all items in $topkPatterns$)

16: Call k-PFPMiner(cPP-List).

Algorithm 2. k-PFPMiner(cPP-List)

1: **for** each item i in cPP-List **do**

2:     Set $tp = \emptyset$ and $X = i$;

3:     **for** each item j that comes after i in the cPP-list **do**

4:         Set $Y = X \cup j$ and $TS^Y = TS^X \cap TS^j$;

5:         Calculate *periodicity* of Y;

6:         **if** $per(TS^Y) \leq dMaxPer$ **then**

7:             Add Y to tp and Y is considered as candidate top-k periodic-frequent itemset;

8:         Check(Y, TS^Y)
            (to check if pattern can make in to top-k periodic-frequent pattern)

9:     $k\text{-}PFPMiner(tp)$

Algorithm 3. Check(X, TS-List)

---

if $per(TS - List) < dMaxPer$ **then**
    Pop the Last pattern and insert X in $topk - patterns$.
    $dMaxPer = max$(periodicity of all items in $topkPatterns$)

Table 2. The complete statistical information of the databases used in our experiments

S.No	Database	Type	Transaction Length (in count)			Items (in count)	Database Size (in count)
			min.	avg.	max.		
1	T10I4D100K	Synthetic	1	10	29	870	1,00,000
2	Retail	Real	2	12	77	16,471	88,162
3	Congestion	Real	1	58	337	1,414	8,928
4	Pollution	Real	11	460	971	1,600	720
5	Kosarak	Real	2	9	2,499	41,270	99,00,00

5.1 Experimental Setup

Our k-PFPMiner algorithm was developed in Python 3.7 and executed on a Giga-byte R282-z94 rack server machine containing two AMD EPIC 7542 CPUs and 600 GB RAM. The operating system of this machine is Ubuntu Server OS 20.04. The experiments have been conducted on both synthetic (**T25I10D10K**) and (**Retail**) and real-world (**Congestion, Pollution**, and **Kosarak**) databases. The T10I10D100K database is a synthetic database generated using the procedure described in [3]. Table 2 presents the statistical information that may be found in the databases mentioned above. Pollution and Congestion may be high-dimensional databases that include extensive transactions.

In the field of intelligent transportation systems, one challenging but important task is monitoring traffic congestion in smart cities. In this regard, the JARTIC situated in Kobe, Japan has established many sensor networks to monitor congestion in several smart cities. *Each transaction in this database takes place every 5 min and includes a timestamp and the identifiers of road segments that have reported traffic jams of more than 300 m.* The time period covered by the data collection is from July 1st to July 31st, 2015.

The Japanese Ministry of the Environment developed the AEROS to tackle air pollution problems. Several air pollution measurement sensors are scattered around Japan as part of this system. Each station collects the data of various air pollutants, say $PM_{2.5}$, NO_2, and O_3, on an hourly basis. For our experiment, we confine to $PM_{2.5}$, since that particular particle size is the primary contributor to the wide variety of cardio-respiratory issues experienced by Japanese citizens. According to Air Quality Index Standards, $PM_{2.5}$ values greater than 16 $\mu g/m^3$ per hour are unsuitable for people.

5.2 Evaluation of Algorithm by Varying only k

Figures 2a, 2b, 2c and 2d show the time consumed at a different number of k values in T10I10D100K, Retail, Congestion and Pollution databases, respectively. It can be observed that an increase in K increases the runtime to find all top-k periodic-frequent patterns being generated at different k values. As k increases, the number of patterns to be mined increases, resulting in time consumption.

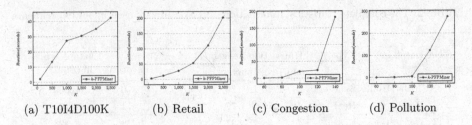

(a) T10I4D100K   (b) Retail   (c) Congestion   (d) Pollution

Fig. 2. Runtime evaluation on various databases by varying k

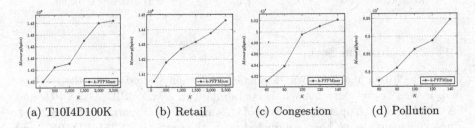

(a) T10I4D100K   (b) Retail   (c) Congestion   (d) Pollution

Fig. 3. Memory evaluation on various databases by varying k

Figures 3a, 3b, 3c and 3d show the memory consumed at a different number of k values in T10I10D200K, Retail, Congestion and Pollution databases, respectively. It can be observed that an increase in K increases the memory to find all top-k periodic-frequent patterns being generated at different k values.

5.3   Scalability Test

In this experiment, we have used the Kosarak database, which is a huge database having 9,90,000 transactions (in count). We have divided this database into five segments, each consisting of 2,00,000 transactions. We have evaluated the performance of k-PFPMiner by adding each successive segment to the ones that came before it. The runtime requirements and memory consumption k-PFPMiner for each segment of the Kosarak database are shown in Fig. 4a and 4b, when $k = 200$. The following are some noteworthy findings that can be derived from these figures: (i) runtime requirements of k-PFPMiner increases almost proportionally as database size grows. (ii) memory requirements of k-PFPMiner where we can observe same as 4a.

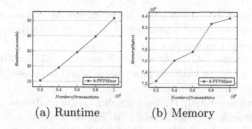

(a) Runtime   (b) Memory

Fig. 4. Scalability of k-PFPMiner

6   Conclusions and Future Work

In this paper, we have proposed an efficient depth-first search algorithm, called top-k Periodic-frequent Pattern Miner (k-PFPMiner), to find all desired patterns in big temporal databases. We have solved the open research problem of setting $maxPer$ and $minSup$ constraints by introducing a novel upper-bound measure named *dynamic maximum periodicity*. With the help of a novel pruning technique, we have reduced the best and worst-case time complexity of identifying whether a pattern is a periodic or aperiodic pattern to O(1) and O(n), respectively. An in-depth examination of the proposed k-PFPMiner approach on four synthetic and real-world databases revealed that its memory consumption and runtime are efficient and highly scalable. As for future work, we will work on discovering top top-k periodic-frequent patterns in uncertain databases.

References

1. Agrawal, R., Imieliński, T., Swami, A.: Mining association rules between sets of items in large databases. In: SIGMOD, pp. 207–216 (1993)
2. Amphawan, K., Lenca, P., Surarerks, A.: Mining top-K periodic-frequent pattern from transactional databases without support threshold. In: Papasratorn, B., Chutimaskul, W., Porkaew, K., Vanijja, V. (eds.) IAIT 2009. CCIS, vol. 55, pp. 18–29. Springer, Heidelberg (2009). https://doi.org/10.1007/978-3-642-10392-6_3
3. Anirudh, A., Kiran, R.U., Reddy, P.K., Kitsuregawa, M.: Memory efficient mining of periodic-frequent patterns in transactional databases. In: 2016 IEEE Symposium Series on Computational Intelligence, pp. 1–8 (2016)
4. Fournier-Viger, P., Yang, P., Kiran, R.U., Ventura, S., Luna, J.M.: Mining local periodic patterns in a discrete sequence. Inf. Sci. **544**, 519–548 (2021)
5. Fournier-Viger, P., Yang, P., Lin, J.C.-W., Kiran, R.U.: Discovering stable periodic-frequent patterns in transactional data. In: Wotawa, F., Friedrich, G., Pill, I., Koitz-Hristov, R., Ali, M. (eds.) IEA/AIE 2019. LNCS (LNAI), vol. 11606, pp. 230–244. Springer, Cham (2019). https://doi.org/10.1007/978-3-030-22999-3_21
6. Kiran, R.U., et al.: Discovering fuzzy periodic-frequent patterns in quantitative temporal databases. In: FUZZ-IEEE 2020, pp. 1–8 (2020)
7. Kiran, R.U., Kitsuregawa, M.: Novel techniques to reduce search space in periodic-frequent pattern mining. In: DASFAA, pp. 377–391 (2014)

8. Kiran, R.U., Shang, H., Toyoda, M., Kitsuregawa, M.: Discovering recurring patterns in time series. In: Proceedings of the 18th International Conference on Extending Database Technology, pp. 97–108 (2015)

9. Kiran, R.U., Venkatesh, J., Toyoda, M., Kitsuregawa, M., Reddy, P.K.: Discovering partial periodic-frequent patterns in a transactional database. J. Syst. Softw. **125**, 170–182 (2017)

10. Luna, J.M., Fournier-Viger, P., Ventura, S.: Frequent itemset mining: a 25 years review. Wiley Interdiscip. Rev. Data Min. Knowl. Discov. **9**(6) (2019)

11. Ravikumar, P., Likhitha, P., Venus Vikranth Raj, B., Uday Kiran, R., Watanobe, Y., Zettsu, K.: Efficient discovery of periodic-frequent patterns in columnar temporal databases. Electronics **10**(12) (2021)

12. Tanbeer, S.K., Ahmed, C.F., Jeong, B.-S., Lee, Y.-K.: Discovering periodic-frequent patterns in transactional databases. In: Theeramunkong, T., Kijsirikul, B., Cercone, N., Ho, T.-B. (eds.) PAKDD 2009. LNCS (LNAI), vol. 5476, pp. 242–253. Springer, Heidelberg (2009). https://doi.org/10.1007/978-3-642-01307-2_24

Hui2Vec: Learning Transaction Embedding Through High Utility Itemsets

Khaled Belghith[1]([✉]), Philippe Fournier-Viger[2], and Jassem Jawadi[3]

[1] Nard Intelligence, Tunis, Tunisia
kb@nard-intelligence.com
[2] College of Computer Science and Software Engineering, Shenzhen University,
Shenzhen, China
philfv@szu.edu.cn
[3] High Institute of Information and Communication Technologies,
University of Carthage, Tunis, Tunisia
jassem.jawadi@istic.ucar.tn
https://nard-intelligence.net

Abstract. Mining frequent itemsets (FIs) in transaction databases is a very popular task in data mining. It helps create meaningful and effective representations for customer transactions which is a key step in the process of transaction classification and clustering. To improve the quality of these representations, previous studies have adapted vector embedding methods to learn transaction embeddings from items and FIs. However, FIs are still a simple pattern type that ignores important information about transactions such as the purchase quantities of items and their unit profits. To address this issue, we propose to learn transaction embeddings from items and high-utility itemsets (HUIs), a more general pattern type. Since HUIs were shown to be more appropriate than FIs for a wide range of applications, we take for hypothesis that transaction embeddings learned from HUIs will be more representative and meaningful. We introduce an unsupervised method, named Hui2Vec, to learn transaction embeddings by combining both singleton items and HUIs. We demonstrate the superior quality of the embedding achieved with the proposed method compared to the embeddings learned from items and FIs on four datasets.

Keywords: Data mining · High-utility itemset mining · Embedding methods · Machine learning

1 Introduction

Transaction classification and clustering are very common data mining problems that have many applications in today's businesses mainly by supporting the decision-making process at different levels [2,12]. When applied to the use case of a supermarket transaction dataset for example, these methods support the decision-making process with an enhanced and insightful customer profiling,

© The Author(s), under exclusive license to Springer Nature Switzerland AG 2022
P. P. Roy et al. (Eds.): BDA 2022 India, LNCS 13773, pp. 211–224, 2022.
https://doi.org/10.1007/978-3-031-24094-2_15

a more accurate and informative market basket analysis, etc. These problems have been well studied in the literature and many different methods have been proposed. Most of them require the use of machine learning techniques such as Support Vector Machines (SVM) and K-means.

Since using machine learning techniques usually requires input data of fixed size whereas there is no obligation for transactions to be of a fixed size, a transformation step is necessary here. That step is commonly referred to as feature extraction, a commonly used mechanism in machine learning. Feature extraction consists of transforming raw data into numerical features that can be processed while preserving the information of the original data set. Feature extraction usually yields better results than applying machine learning directly to the raw data.

To generate useful features for transaction data, a promising approach is to use frequent itemsets (FIs) as features [1,2]. A frequent itemset is a set of values that frequently appears in transactions. After extracting FIs, each transaction can be converted into a vector of binary values, each indicating whether a specific frequent itemset is present in the transaction. Other more effective approaches introduced discriminative measures to handle the problem of high-dimensionality and data sparsity on datasets with a large number of FIs [3,4]. Using discriminative measures helps extract only significant FIs but makes the feature extraction process supervised. These supervised FI-based approaches rely on the availability of labeled data. However, labels are not always provided and can be difficult to produce. The result is a representation tailored to a specific mining task that cannot be directly transferred to other tasks [5].

To cope with the limitations of FI-based and supervised FI-based approaches, Trans2Vec [5] was proposed as a new unsupervised embedding method for learning low-dimensional representations of transactions. Trans2Vec is inspired from embedding methods widely used in machine learning mainly in text and natural language processing [14–16]. It learns transaction embeddings from information of both singleton items and FIs. Trans2Vec was shown to outperform classical FI-based methods on several benchmark datasets [5].

The aforementioned FIs-based approaches are useful but an important limitation remains. It is that frequent itemset mining (FIM) algorithms for extracting FIs rely on a simplified transaction format where each item appears or not in each transaction. But in real-life, transactions data contain additional information such as the purchase quantities of items and their unit profits. To find itemsets that are profitable rather than frequent, FIM was generalized as high-Utility Itemset Mining (HUIM) [7–9]. The aim is to discover itemsets that have a high utility (importance) in transactions. The utility of an itemset can be measured according to several criteria such as the profit, frequency or weight. High utility itemsets (HUIs) and their variations were shown to be more appropriate than FIs for a wide range of applications such as market basket analysis, clickstream analysis [11], biomedicine [12], cross-marketing and mobile commerce [10]. Nonetheless, to our best knowledge, there are still no studies that have attempted to benefit from HUIs to create better transaction embeddings than using FIs.

In this paper, we present a new method to learn transaction embeddings through high utility itemsets, called Hui2Vec. It extends Trans2Vec to no longer operate at the level of FIs but at the level of HUIs. To extract HUIs, we adapt the EFficient high-utility Itemset Mining (EFIM) [6] algorithm, which was shown to be two to three times faster than many other state-of-the-art HUIM algorithms such as HUP-Miner [9] and FHM [8]. Th three main contributions of this study are:

1. We propose **Hui2Vec**, an unsupervised method, to learn low-dimensional continuous representations for transaction data.
2. We propose two models in **Hui2Vec**, which learn transaction embeddings from information of both singleton items and HUIs. We use EFIM [6], a very efficient state of the art algorithm to mine high utility itemsets.
3. We demonstrate that **Hui2Vec** is useful for transaction classification, as it achieves better results than Trans2Vec on several datasets.

2   Related Work

The proposed method, called Hui2Vec, is related to high-utility itemset mining (HUIM). The problem of HUIM has been widely studied in the literature. The traditional approach to HUIM is to discover high-utility itemsets in two phases using the Transaction-Weighted-Downward Closure model [20]. In the first phase, a set of candidate high-utility itemsets is generated by overestimating their utility. In the second phase, the database is scanned to calculate the exact utilities of candidates and filter low-utility itemsets. Many algorithms follow this two-phase approach such as IHUP [7], Two-Phase [20], BAHUI [21], UP-Growth+ [10] and MU-Growth [22].

To overcome the problem of efficiency caused by the two very time-consuming phases, recent algorithms mine high-utility itemsets in a single phase. The first algorithms of this type are HUI-Miner and d^2HUP [19]. They avoid the problem of candidate generation and were reported to be 10 to 100 times faster than legacy two-phase approaches [19]. HUP-Miner [9] and FHM [8], two improved versions of HUI-Miner were proposed and shown to be 6 times faster than HUI-Miner. Despite all these improvements, high utility itemset mining approaches remain very expensive in time and memory consumption [8, 9, 19].

To go one step further into improving the efficiency of HUIM methods, EFIM, for Efficient high-utility Itemset Mining, a new one-phase algorithm was proposed [6]. The new approach introduced several new ideas that helped improve the performance both in memory consumption and execution time. EFIM design is based on the principle that all operations on each itemset should be performed in linear time. First, it introduced two efficient techniques to reduce the cost of database scans named High-utility Database Projection (HDP) and High-utility Transaction Merging (HTM). Second, it introduced two new upper-bounds on the utility of itemsets named the revised sub-tree utility and local utility to more effectively prune the search space. Finally, a novel array-based utility counting technique named Fast Utility Counting (FAC) was used to calculate these upper-bounds in linear time and space for all extensions of an itemset. The authors conducted extensive experiments to compare EFIM to other state-of-the-art algorithms [6]. The results showed that EFIM is always performing better than other algorithms both in execution time and in memory consumption [6]. This is why we choose EFIM to extract high utility itemsets in the preparation step in Hui2Vec before we start learning the embedding.

The novel Hui2Vec approach is also related to vector embedding methods. These are learned representations of the data in the form of numerical vectors. The performance of machine learning methods crucially depends on the quality of these vector representations [23]. A wide range of methods for generating vector embeddings have been studied [23]. Among them, Word2Vec [14] is a well-known adaptation to learn embedding vectors for words in text.

Word2Vec inspired approaches to learn documents in text [15,16] and approaches to learn node embeddings in graphs [24,27,28]. DeepWalk [27] and node2vec [28] are based on interpreting a sequence of nodes seen on random walks within a graph as if they were words appearing together in a sentence. Recent approaches are extending vector embeddings to learn graph representations [23,26,29,30]. Trans2Vec.[5] is an adaptation of vector embedding to learn transaction embeddings. Trans2Vec is fully unsupervised and learns a transaction embedding from both items and FIs. In this paper, we present Hui2Vec, a new approach that extends Trans2Vec to learn transaction embedding from items and HUIs.

3   Framework

3.1   Problem Definition

We follow the same notations as in [2] and in [5]. Given a set of items $I = \{i_1, i_2, ..., i_M\}$, a transaction dataset $D = \{T_1, T_2, ..., T_N\}$ is a set of transactions where each transaction T_i is a set of distinct items (i.e., $T_i \subseteq I$).

Our goal is to learn a mapping function $f : D \to \mathbb{R}^d$ such that every transaction $T_i \in D$ is mapped to a d-dimensional continuous vector. The mapping needs to capture the similarity among the transactions in D, in the sense that T_i and T_j are similar if $f(T_i)$ and $f(T_j)$ are close to each other in the vector space, and vice versa. The matrix $X = [f(T_1), f(T_2), ..., f(T_N)]$ then contains feature vectors of transactions, which can be direct inputs for many traditional machine learning and data mining tasks, particularly classification.

3.2   Learning Transaction Embeddings Based on Items

We follow the same methodology as in Trans2Vec [5] with a part of HUIM embedding coming from singleton items directly. We adapt the Vector-Distributed Bag-of-Words (PV-DBOW) [15] where each transaction is a document and items are the words. Given a transaction T_t, and a set of items $I(T_t) = \{i_1, i_2, ..., i_k\}$

contained in T_t, we need to learn the representation of T_t by maximizing the log probability of predicting i_1, i_2, ..., i_k appearing in T_t :

$$max \Sigma_{j=1}^{k} log Pr(i_j \mid T_t) \tag{1}$$

$Pr(i_j \mid T_t)$ is defined by a softmax function:

$$Pr(i_j \mid T_t) = \frac{exp(g(i_j).f(T_t))}{\Sigma_{i' \in I} exp(g(i').f(T_t))} \tag{2}$$

where $g(i_j) \in R^d$ is the embedding vector of item i_j and $f(T_t) \in R^d$ is the embedding vector of the transaction T_t.

We follow the negative sampling technique in Word2Vec [14] to avoid iterating on all items in I and instead select a small number of items not contained in the transaction T_t. The problem become in the form of minimizing this binary objective function of logistic regression:

$$\mathcal{O}_1 = -[log\sigma(g(i_j).f(T_t)) + \sum_{n=1}^{k} \mathbb{E}_{i^n \sim P(i)} log\sigma(-g(i^n).f(T_t))] \tag{3}$$

where $\sigma(x) = \frac{1}{1+e^{-x}}$ is a sigmoid function, $P(i)$ is the negative item collection, i^n is a negative item taken from $P(i)$ for K times, and $g(i^n) \in \mathbb{R}^d$ is the embedding vector of i^n.

The derived gradients resulting from minimizing the objective function in Eq. 3 using stochastic gradient descent (SGD) are the following:

$$\frac{\partial \mathcal{O}_1}{\partial g(i^n)} = -\sigma(g(i^n).f(T_t) - \mathbb{I}_{ij}[i^n]).f(T_t)$$

$$\frac{\partial \mathcal{O}_1}{\partial f(T^t)} = -\sum_{n=0}^{k} \sigma(g(i^n).f(T_t) - \mathbb{I}_{ij}[i^n]).g(i^n) \tag{4}$$

where $\mathbb{I}_{ij}[i^n]$ is a function indicating whether i^n is an item i_j (i.e., the negative item is contained in the transaction T_t), and if $n = 0$, then $i^n = i_j$.

3.3 Learning Transaction Embedding Based on High Utility Itemsets

As mentioned in Sect. 1, we take for hypothesis that using HUIs instead of FIs would be more effective since HUIs would capture more information and are generally more appropriate for a wide range of applications such as market basket analysis, clickstream analysis, and bio-medicine. With Hui2Vec, we learn transaction embeddings based on HUIs instead of FIs, as was the case with Trans2Vec. We believe that the transaction representation learned with Hui2Vec will be more meaningful and discriminative than with Trans2Vec, especially in applications where HUIs are more informative than FIs.

Table 1. A sample transaction database

TID	Transaction
T_1	(a,1) (c,1) (d,1)
T_2	(a,2) (c,6) (e,2) (g,5)
T_3	(a,1) (b,2) (c,1) (d,6) (e,1) (f,5)
T_4	(b,4) (c,3) (d,3) (e,1)
T_5	(b,2) (c,2) (e,1) (g,2)

The adaptation we make here to learn embeddings based on HUIs is very similar to the work made in Trans2Vec to learn Transaction Embeddings Based on Frequent Itemsets [5]. We follow the same notations in [6] and define a high utility itemset as follows. Let I be a finite set of items (symbols). An itemset X is a finite set of items such that $X \subseteq I$. A transaction database is a multiset of transactions $D = \{T_1, T_2, ..., T_N\}$ such that for each transaction T_c, $T_c \subseteq I$ and T_c has a unique identifier c called its TID (Transaction ID). Each item $i \in I$ is associated with a positive number $p(i)$, called its external utility. The external utility of an item represents its relative importance to the user. Every item i appearing in a transaction T_c has a positive number $q(i, T_c)$, called its internal utility. In the context of market basket analysis, the external utility typically represents an item's unit profit (for example, the sale of 1 unit of bread yields a 1 dollar profit), while the internal utility represents the purchase quantity of an item in a transaction (for example, a customer has bought 2 units of bread).

Table 2. External utility values

Item	a	b	c	d	e	f	g
Profit	5	2	1	2	3	1	1

Example. Consider the example transaction dataset in Table 1 with five transactions $(T_1, T_2...T_5)$. Transaction $T2$ indicates that items a, c, e and g appear in this transaction with an internal utility (e.g. purchase quantity) of respectively 2, 6, 2 and 5.

Table 2 indicates that the external utility (e.g. unit profit) of these items are respectively 5, 1, 3 and 1. The utility of an itemset X is denoted as $u(X)$ and defined as $u(X, T_c) = \sum_{i \in X} u(i, T_c)$ if $X \subseteq T_c$. Otherwise $u(X, T_c) = 0$. The utility of an itemset X is denoted $u(X)$ and defined as $u(X) = \sum_{T_c \in g(X)} u(i, T_c)$, where $g(X)$ is the set of transactions containing X. It represents the profit generated by the sale of the itemset X. For example, the utility of the item a in the transaction T_2 is $u(a, T_2) = 5 \times 2 = 10$. The utility of the itemset $\{a, c\}$ in T_2 is $u(\{a, c\}, T_2) = u(a, T_2) + u(c, T_2) = 5 \times 2 + 1 \times 6 = 16$. The utility of the itemset

$\{a, c\}$ is $u(\{a, c\}) = u(a)+u(c) = u(a, T_0)+ u(a, T_2) + u(a, T_3) + u(c, T_0) +$
$u(c, T_2) + u(c, T_3) = 5 + 5 + 10 + 1 + 1 + 6 = 28$.

Table 3. High utility itemset for $minutil = 30$

HUI	Itemset	Utility
$X1$	$\{b, d\}$	30
$X2$	$\{a, c, e\}$	31
$X3$	$\{b, c, d\}$	34
$X4$	$\{b, c, e\}$	31
$X5$	$\{b, d, e\}$	36
$X6$	$\{b, c, d, e\}$	40
$X7$	$\{a, b, c, d, e, f\}$	30

The problem of high-utility itemset mining is to discover all high-utility itemsets, given a threshold $minutil$ set by the user. If $minutil = 30$, the high-utility
itemsets or HUIs in the database of the running example are shown in Table 3.
Each Transaction now can now be represented by a set of HUIs, as shown in
Table 4.

Table 4. Each transaction represented by a set of HUIs

TID	HUIs
T_1	$\{\}$
T_2	$\{X_2\}$
T_3	$\{X_1, X_2, X_3, X_4, X_5, X_6, X_7\}$
T_4	$\{X_1, X_3, X_4, X_5, X_6\}$
T_5	$\{X_4\}$

Following the same procedure in the previous section, given a set of HUIs
$F(T_t) = \{X_1, X_2, ..., X_l\}$ contained in a transaction T_t based on its HUIs is
defined as follows:

$$\mathcal{O}_2 = -[log\sigma(h(X_j).f(T_t)) + \sum_{n=1}^{k} \mathbb{E}_{X^n \sim P(X)} log\sigma(-h(X^n).f(T_t))] \quad (5)$$

where $h(X_j) \in \mathbb{R}^d$ is the embedding vector of the frequent itemset $X_j \in$
$F(T_t)$, $P(X)$ is the negative item collection (i.e., a small set of random HUIs
which are not contained in T_t), X^n is a negative item taken from $P(X)$ for K
times, and $h(X^n) \in \mathbb{R}^d$ is the embedding vector of X^n. Here again the objective
function in Eq. 5 is minimized using SGD.

3.4   Hui2Vec Methods to Learn Transaction Embeddings

When representing each transaction by a set of HUIs, we can face a situation where a Transaction T_t does not contain any HUIs. This is the case for example with transaction T_1 in Table 4. In this case, learning a useful embedding is very difficult. To overcome this problem, a good idea could be to combine information of both items and HUIs to learn embedding vectors for transactions. Here again, we follow the same model implemented in Trans2Vec [5].

Individual-Training Model to learn Transaction Embeddings. The method followed here, as illustrated in Fig. 1, learns an embedding vector $f_1(T_t)$ for the transaction T_t based on its items and an embedding vector $f_2(T_t)$ based on its HUIs. The average of the two embedding vectors is taken as final embedding vector for the transaction $f(T_t) = \frac{f_1(T_t)+f_2(T_t)}{2}$.

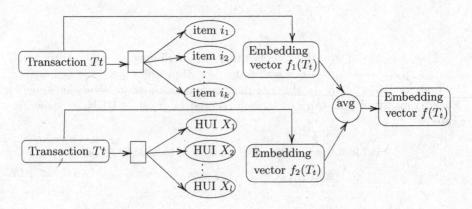

Fig. 1. The individual training model of Hui2Vec

Joint-Training Model to Learn Transaction Embeddings. In the designed method, shown in Fig. 2, both information from items and HUIs are combined. More precisely, given a Transaction T_t, the objective function that needs to be minimized is the following:

$$\mathcal{O} = -[\sum_{i_j \in I(T_t)} logPr(i_j \mid T_t) + \sum_{X_j \in F(T_t)} logPr(X_j \mid T_t)] \qquad (6)$$

where $I(T_t)$ is the set of singleton items in T_t and $f(T_t)$ is the set high-utility items or HUIs contained in T_t. The Eq. 6 can be simplified to this form:

$$\mathcal{O} = - \sum_{p_j \in I(T_t) \cup F(T_t)} logPr(p_j \mid T_t), \qquad (7)$$

where $p_j \subseteq T_t$ is an item or a high utility item. Usually, we refer to p_j as a pattern.

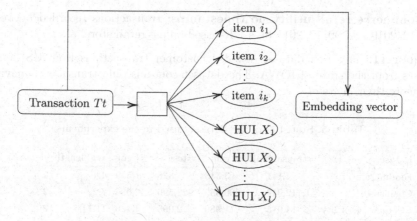

Fig. 2. The Joint-Training Model of Hui2Vec

The objective function here becomes the following:

$$\mathcal{O} = -[log\sigma q(p_j).f(T_t) + \sum_{n=1}^{k} \mathbb{E}_{p^n \sim P(p)} log\sigma(-q(p^n).f(T_t))] \quad (8)$$

where $q(p_j) \in \mathbb{R}^d$ is the embedding vector of the pattern $p_j \in I(T_t) \cup F(T_t)$, $P(p)$ is the negative pattern collection (i.e., random patterns which are not contained in T_t), p^n is a negative pattern drawn from $P(p)$ for K times, and $q(p^n) \in \mathbb{R}^d$ is the embedding vector of p^n. The minimization is here also made using SGD. After the learning process is completed, a transaction T_i will be close to a transaction T_j if they have similar items and HUIs.

4   Experiments

We conduct extensive experiments on real-world transaction datasets to evaluate Hui2Vec's performance for transaction classification.

4.1   Datasets

The statistics of the datasets used in the experiments are shown in Table 5. The columns 2 to 5 indicate the total number of transactions, the number of transactions for training and testing, the number of distinct items, the average transaction length, and the number of classes. All these datasets are taken from SPMF [17], an Open-Source Data Mining Library.

Foodmart_utility: dataset of customer transactions from a retail store, obtained and transformed from SQL-Server 2000.

Chainstore_utility: dataset of customer transactions from a major grocery store chain in California, USA, containing 1,112,949 transactions and 46,086 items, obtained and transformed from NU-Mine Bench.

ECommerce_retail_utility_no_timestamps: transactions recorded between 01/12/2010 and 09/12/2011 by a UK-based online retail store.

Liquor_11: This is a dataset of 9,284 customer transactions from US liquor stores from the state of IOWA. The dataset contains only transactions having no more than 11 items.

Table 5. Statistics of the datasets used in the experiments

Dataset	#transactions	#train	#test	#items	avg. length	#classes
Foodmart_utility	4,141	3,313	828	1,559	4.4	2
Chainstore	1,112,949	890,359	222,590	46,086	7.2	2
ECommerce_utility	14,975	11,980	2,995	3,468	11.71	2
Liquor_11	9,284	7,427	1,857	2,626	2.7	3

4.2   Implemented Models

As mentioned in the previous section, we implemented two different models for the proposed Hui2Vec method to learn transaction embeddings based on items and HUIs. The first model **Hui2Vec-IND** learns transaction embeddings from items and HUIs separately and then take the average. The second model **Hui2Vec-JOI** learns transaction embeddings from items and HUIs simultaneously. To find high utility itemsets in a dataset, we use EFIM's [6] implementation from the SPMF library [17].

To validate the efficiency of the designed Hui2Vec models, we compare them with the two corresponding models from the Trans2Vec approach [5], **Trans2Vec-IND** and **Trans2Vec-JOI**. Again, Trans2Vec models learn transaction embeddings from items and FIs. **Trans2Vec-IND** behaves in a way very similar to **Hui2Vec-IND** but uses HUIs instead of FIs. The same with **Trans2Vec-JOI** and **Hui2Vec-JOI**.

4.3   Evaluation Metrics

To evaluate the proposed models, we follow the same methodology as in [5]. Once embeddings of transactions are learned, we feed them to an SVM with linear kernel [31] to classify the transaction labels. We use the linear-kernel SVM (a simple classifier) and do not tune the parameter C of SVM (here, we fix $C = 1$) since our focus is on the transaction embedding learning, not on a classifier. Each dataset is randomly shuffled and split into the training and test sets as shown in Table 1. All methods are applied to the same training and test sets. We repeat the classification process on each dataset ten times and report the average classification accuracy and the average F1-macro score. We do not report the standard deviation since all methods are very stable (their standard deviations are less than 10^{-2}).

4.4   Parameter Settings

Our method Hui2Vec has two important parameters: the high-utility threshold *minutil* for extracting HUIs and the embedding dimension d. Trans2Vec the method we are comparing our model to has also two important parameters: the minimum support threshold δ for extracting FIs and the embedding dimension d. Since we are developing a fully unsupervised approach, all these values are assigned without using transaction labels. We set $d = 128$ which is a common value used in embedding methods [5]. We set δ and *minutil* following the *elbow method* [5,32].

Table 6. Experiments results - Comparison between *Trans2Vec* and *Hui2Vec* in Accuracy (AC) and F1-macro (F1) on the four datasets.

Method	Foodmart_utility		Chainstore		Ecommerce_utility		Liquor_11	
	AC	F1	AC	F1	AC	F1	AC	F1
Trans2Vec IND	50.1	49.77	43.09	42.81	50.27	47.50	43.67	43.50
Trans2Vec JOI	67.06	66.89	52.90	52.74	82.52	82.36	81.15	81.45
Hui2Vec IND	92.9	92.55	80.62	79.55	94.76	94.42	93.34	93.57
Hui2Vec JOI	**93.58**	**93.33**	**81.43**	**80.46**	**95.43**	**95.11**	**94.25**	**94.83**

4.5   Results and Discussion

Results are shown in Table 6. Two main observations are made. First, we can see that the two models of the proposed Hui2Vec method are clearly providing better classification on all datasets compared to the two other models from Trans2Vec. **Hui2Vec-JOI** improved accuracy over **Trans2Vec-JOI** on all datasets by large margins (achieving 15–52 % gains). This demonstrates that the proposal of incorporating information of both singleton items and HUIs into the transaction embedding learning is a better strategy than learning transaction embeddings from combining items and FIs. This is reasonable because Trans2Vec relies on FIs and thus can only capture frequent relationships between items, while Hui2Vec considers not only relationships between items but also the utility information (unit profit and quantities). This is interesting as utility information is generally available in real transaction datasets such as those used in the experiments. Hence, it is expected that in pratice, no additional effort is needed for users to collect data to apply Hui2Vec compared to Trans2Vec.

Second, we find that **Hui2Vec-JOI** produces better results than **Hui2Vec-IND** on all datasets. This confirms the idea that transaction embeddings learned from items and HUIs simultaneously are more meaningful and discriminative since they can capture different latent relationships of transactions simultaneously. Similarly, it is also observed that **Trans2Vec-JOI** always perform better than **Trans2Vec-IND** on all datasets.

On overall, experiments show good classification performance using the proposed Hui2Vec method to learn transaction embeddings.

5   Conclusion

In this paper, we demonstrated that learning transaction embeddings from both singleton items and HUIs produces more meaningful and discriminative representations than learning from items with FIs. We presented Hui2Vec, an unsupervised method for learning transaction embeddings from combining both singleton items and HUIs. We implemented two different models of the Hui2Vec approach. Experiments were carried out on four datasets, which showed that the Hui2Vec models significantly outperform the Trans2Vec models [5], which learns from items and FIs, in both accuracy and F1-macro scores.

As an extension to this work, we plan on studying the impact of parameters on the method (the high-utility threshold *minutil* for extracting HUIs and the embedding dimension d) on the quality of the embeddings produced. Another extension is to investigate the quality of the embeddings for the transaction clustering task.

Acknowledgment. Authors would like to thank the authors of Trans2Vec [5] for providing their source code. This work is partially supported by NARD Intelligence[1].

References

1. Cheng, H., Yan, X., Han, J., Hsu, C.-W.: Discriminative frequent pattern analysis for effective classification. In: ICDE 2007, pp. 716–725 (2007)
2. Fournier-Viger, P., Lin, J.C.-W., Vo, B., Chi, T.T., Zhang, J., Le, H.B.: A survey of itemset mining. Wiley Interdiscip. Data Min. Knowl. Discov. **7**(4), e1207 (2017)
3. He, Z., Feiyang, G., Zhao, C., Liu, X., Jun, W., Wang, J.: Conditional discriminative pattern mining: concepts and algorithms. Inf. Sci. **375**, 1–15 (2017)
4. Kameya, Y., Sato, T.: RP-growth, Top-k mining of relevant patterns with minimum support raising. In: SIAM International Conference on Data Mining 2012, pp. 816–827 (2012)
5. Nguyen, D., Nguyen, T.D., Luo, W., Venkatesh, S.: Trans2Vec: learning transaction embedding via items and frequent itemsets. In: Phung, D., Tseng, V.S., Webb, G.I., Ho, B., Ganji, M., Rashidi, L. (eds.) PAKDD 2018. LNCS (LNAI), vol. 10939, pp. 361–372. Springer, Cham (2018). https://doi.org/10.1007/978-3-319-93040-4_29
6. Zida, S., Fournier-Viger, P., Chun-Wei Lin, J., Wu, C.W., Tseng, V.S.: EFIM: a fast and memory efficient algorithm for high-utility itemset mining. Knowl. Inf. Syst. **51**(2), 595–625 (2017)
7. Ahmed, C.F., Tanbeer, S.K., Jeong, B.S., Lee, Y.K.: Efficient tree structures for high-utility pattern mining in incremental databases. IEEE Trans. Knowl. Data Eng. **21**(12), 1708–1721 (2009)
8. Fournier-Viger, P., Wu, C.-W., Tseng, V.S.: Novel concise representations of high utility itemsets using generator patterns. In: Luo, X., Yu, J.X., Li, Z. (eds.) ADMA 2014. LNCS (LNAI), vol. 8933, pp. 30–43. Springer, Cham (2014). https://doi.org/10.1007/978-3-319-14717-8_3
9. Krishnamoorthy, S.: Pruning strategies for mining high utility itemsets. Expert Syst. Appl. **42**(5), 2371–2381 (2015)
10. Tseng, V.S., Shie, B.E., Wu, C.W., Yu, P.S.: Efficient algorithms for mining high utility itemsets from transactional databases. IEEE Trans. Knowl. Data Eng. **25**(8), 1772–1786 (2013)

11. Thilagu, M., Nadarajan, R.: Effciently mining of effective web traversal patterns with average utility. In: Proceedings of the International Conference on Communication, Computing, and Security, pp. 444–451. CRC Press (2012)
12. Fournier-Viger, P., Lin, J.C.-W., Nkambou, R., Vo, B., Tseng, V.S. (eds.): High-Utility Pattern Mining. SBD, vol. 51. Springer, Cham (2019). https://doi.org/10.1007/978-3-030-04921-8
13. Liu, Y., Cheng, C., Tseng, V.S.: Mining differential top-k co-expression patterns from time course comparative gene expression datasets. BMC Bioinform. 14(230) (2013)
14. Mikolov, T., Sutskever, I., Chen, K., Corrado, G.S., Dean, J.: Distributed representations of words and phrases and their compositionality. In: NIPS 2013, pp. 3111–3119 (2013)
15. Le, Q., Mikolov, T.: Distributed representations of sentences and documents. In: ICML 2014, pp. 1188–1196 (2014)
16. Chen, M.: Efficient vector representation for documents through corruption. In: ICLR 2017 (2017)
17. Fournier-Viger, P., Gomariz, A., Gueniche, T., Soltani, A., Wu, C.W., Tseng, V.S.: SPMF: a java open-source pattern mining library. J. Mach. Learn. Res. 15, 3389–3393 (2014)
18. Lan, G.C., Hong, T.P., Tseng, V.S.: An efficient projection-based indexing approach for mining high utility itemsets. Knowl. Inf. Syst. 38(1), 85–107 (2014)
19. Liu, J., Wang, K., Fung, B.: Direct discovery of high utility itemsets without candidate generation. In: Proceedings of the 12th IEEE International Conference on Data Mining, IEEE, Brussels, Belgium, December 2012, p. 984989 (2012)
20. Liu, Y., Liao, W., Choudhary, A.: A two-phase algorithm for fast discovery of high utility itemsets. In: Ho, T.B., Cheung, D., Liu, H. (eds.) PAKDD 2005. LNCS (LNAI), vol. 3518, pp. 689–695. Springer, Heidelberg (2005). https://doi.org/10.1007/11430919_79
21. Song, W., Liu, Y., Li, J.: BAHUI: fast and memory efficient mining of high utility itemsets based on bitmap. Proc. Int. J. Data Warehous. Min. 10(1), 1–15 (2014)
22. Yun, U., Ryang, H., Ryu, K.H.: High utility itemset mining with techniques for reducing overestimated utilities and pruning candidates. Expert Syst. Appl. 41(8), 3861–3878 (2014)
23. Grohe, M.: Word2vec, Node2vec, Graph2vec, X2vec: towards a theory of vector embeddings of structured data. In: ACM SIGMOD-SIGACT-SIGAI Symposium on Principles of Database Systems, PODS 2020, pp. 1–16 (2020)
24. Luo, J., Xiao, S., Jiang, S.: Ripple2Vec: node embedding with ripple distance of structures. Data Sci. Eng. 7, 156–174 (2022)
25. Cao, S., Lu, W., Xu, Q.: GraRep. Learning graph representations with global structural information. In: Proceedings of the 24th ACM International on Conference on Information and Knowledge Management, pp. 891–900 (2015)
26. Ou, M., Cui, P., Pei, J., Zhang, Z., Zhu, W.: Asymmetric transitivity preserving graph embedding. In: Proceedings of the 22nd ACM SIGKDD International Conference on Knowledge Discovery and Data Mining, pp. 1105–1114 (2016)
27. Perozzi, B., Al-Rfou, R., Skiena, S.: Deepwalk: online learning of social representations. In: Proceedings of the 20th ACM SIGKDD International Conference on Knowledge Discovery and Data Mining, pp. 701–710 (2014)
28. Grover, A., Leskovec., J.: Node2Vec: scalable feature learning for networks. In: Krishnapuram, B.B., Shah, M., Smola, A.J., Aggarwal, C.C., Shen, D., Rastogi, R. (eds.), Proceedings of the 22nd ACM SIGKDD International Conference on Knowledge Discovery and Data Mining, pp. 855–864 (2016)

29. Narayanan, A., Chandramohan, M., Venkatesan, R., Chen, L., Liu, Y., Jaiswal., S.: Graph2Vec: learning distributed representations of graphs. ArXiv (CoRR), arXiv:1707.05005 [cs.AI] (2017)
30. Pan, S., Hu, R., Long, G., Jiang, J., Yao, L., Zhang, C.: Adversarially regularized graph autoencoder for graph embedding. In: Proceedings of the 27th International Joint Conference on Artificial Intelligence, IJCAI 2018. pp. 2609–2615 (2018)
31. Chang, C.-C., Lin, C.-J.: LIBSVM: a library for support vector machines. ACM Trans. Intell. Syst. Technol. **2**(3), 1–27 (2011)
32. Rousseau, F., Kiagias, E., Vazirgiannis, M.: Text categorization as a graph classification problem. In: ACL 2015, pp. 1702–1712 (2015)

Predictive Analytics in Agriculture

A Data-Driven, Farmer-Oriented Agricultural Crop Recommendation Engine (ACRE)

Rohit Patel, Inavamsi Enaganti, Mayank Ratan Bhardwaj, and Y. Narahari[✉]

Indian Institute of Science, Bengaluru, India
{rohitpatel,inavamsie,mayankb,narahari}@iisc.ac.in

Abstract. Agriculture has a significant role to play in any emerging economy and provides the source of income and employment for a large portion of the population. A key challenge faced by small and marginal farmers is to determine which crops to grow to maximize their utililty. With a wrong choice of crops, farmers could end up with sub-optimal yields and low, and possibly even loss of revenue. This work seeks to design and develop ACRE (Agricultural Crop Recommendation Engine), a tool that provides a scientific method to choose a crop or a portfolio of crops, to maximize the utility to the farmer. ACRE uses available data such as soil characteristics, weather conditions, and historical yield data, and uses state-of-the-art machine learning/deep learning models to compute an estimated utility to the farmer. The main idea of ACRE is to generate several recommendations of portfolios of crops, with a ranking of portfolios based on the Sharpe ratio, a popular risk metric in financial investments. We use publicly available data from *agmarknet* portal in India to perform several thought experiments with ACRE. ACRE provides a rigorous, data-driven backend for designing farmer-friendly mobile apps for assisting farmers in choosing crops (This work was supported by the National Bank for Agriculture and Rural Development (NABARD), Government of India, through a research grant).

Keywords: Crop recommendation · Crop portfolio · Yield estimation · Machine learning · Deep learning · Sharpe ratio · Risk analysis

1 Introduction

There are numerous challenges faced by small and marginal farmers in all countries where agriculture is a major source of livelihood and employment for the population. In India, for example, agriculture is the most important source of income for around 58% of the inhabitants of the nation [5,14]. According to the 2016–17 Economic Survey in India, a farmer's average monthly income in 17 states is around a meagre Rs. 1700 (approx. US$ 20), resulting in farmer distress and even suicides. There is an increasing trend of conversion of agricultural

© The Author(s), under exclusive license to Springer Nature Switzerland AG 2022
P. P. Roy et al. (Eds.): BDA 2022 India, LNCS 13773, pp. 227–248, 2022.
https://doi.org/10.1007/978-3-031-24094-2_16

land for non-agricultural purposes. Furthermore, 48 percent of farmers are not in favour of their children taking up agriculture as their profession, preferring to live in cities.

1.1   Motivation for ACRE

A crucial reason for the struggles faced by small and marginal farmers is that they frequently make poor crop-selection decisions. Crop selection is an extremely important decision for farmers and there are numerous factors to be taken into account while choosing a crop or a combination of crops.

Factors Influencing Crop Selection. The handbook by Chandra Shekara et al. [22] has enunciated a number of factors influencing the choice of crops. We provide a brief summary here. The factors include [22, 25]:

1. **Climatic factors:** Temperature, rainfall, sun shine hours, relative humidity, wind velocity, wind direction, seasons, and agro-ecological situations.
2. **Soil factors:** Soil type, pH value, and soil fertility.
3. **Water availability:** Tanks, wells, dams, ground water, rainfall, water quality, water suitability, resources for lifting water, and availability of micro-irrigation systems.
4. **Cropping system options:** Inter-cropping, mixed cropping, multi-storeyed cropping, relay cropping, and crop rotation.
5. **Socio-Cultural factors:** Traditional best practices; farmer beliefs and superstition; opinion of family members, neighbours, and friends.
6. **Expected profit:** Profit expected from selling the crops post harvest. There is inherent risk in the market conditions.
7. **Expected yield:** Estimated yield of a crop during harvest. There is inherent risk due to crop damage or loss.
8. **Economic conditions:** Land holding of the farmer; financial resources (including credit availability); labour availability and affordability; availability and affordability of farm mechanization.
9. **Technological factors:** Access to modern technology; feasibility of technology options.
10. **Market demand and access:** Demand for the crop in neighbouring areas; Accessibility to markets and real-time market information.
11. **Government policies and schemes:** Availability incentives, schemes, accessibility to Government call centres, access to farmer producer organisations (FPOs), etc.
12. **Agricultural inputs:** Availability and access to high quality inputs (seeds, fertilizers, pesticides, etc.).
13. **Post harvest storage and processing:** Access to storage facilities, food processing units, technologies for adding value to crops, etc.

It is difficult to expect small and marginal farmers to weigh-in all the factors described above in an informed or algorithmic way and choose the best crops. Lack of knowledge, lack of awareness, and interference of local intermediaries are major obstacles for small and marginal farmers to choose the best portfolio of crops. This is the major motivation for us to develop ACRE (Agricultural Crop Recommendation Engine), a tool that provides a scientific method for choosing a crop or a portfolio of crops to maximize the utility to the farmer. ACRE provides a powerful back-end (computational engine) that embeds rigorous algorithms to provide decision support to the farmer in crop selection. To use ACRE effectively, there is a need to provide a farmer-friendly interface (for example a mobile app) that gathers only essential information from the farmer and provides a crop portfolio recommendation using ACRE. In this paper, our focus is on the computational engine. Clearly, the farmer should be required to provide minimal information with a simple user interface and should not worry at all about the computational engine.

1.2 Contributions and Outline

ACRE works with information such as location of the farmer, feasible subsets of crops, season (Kharif, Rabi, whole year), utility functions of farmers, irrigation facilities, crop rotation cycle, area (in hectares), human labor cost, etc. Some of these are provided by the farmer and ACRE infers the rest of the information. ACRE either sources or infers the parameters such as temperature, rainfall, sunlight, humidity, soil type, and soil nutrients. The computational engine in ACRE computes estimates of yield and cost, as well as, the standard deviations of these estimates. The system then computes an appropriate farmer utility for each crop and provides individual crop recommendations for Kharif and Rabi seasons, based on these utilities. The system also provides inter-crop recommendations, considering several portfolios of crops that could be grown in a given season. A technical novelty of ACRE the use of Sharpe ratio [20, 21], a portfolio investment metric widely used in financial portfolio selection, for evaluating crop portfolios.

One of the significant components of ACRE is crop yield estimation. Crop yield is a complex variable influenced by several factors, including genotype, environment, and interactions. For accurate yield prediction, a fundamental understanding of the functional relationship between yield and these interaction components is required [9]. For yield prediction, we use an ensemble technique consisting of standard machine learning and deep learning regression models. Previous efforts reported in the state-of-the-art literature compute an average yield for different crops. They do not compute the variance of yield which is a good indicator of the risk involved in selecting a given crop. To accurately capture the variance in yield, we use the triangular distribution to capture yield data by considering three parameters: mode, maximum, and minimum values of the crop yield.

Section 2 provides a review of relevant work on crop recommendation and crop yield estimation. Section 3 provides an overview of the Sharpe ratio and brings out its relevance for agriculture crop portfolio selection. Section 4 provides details of the public datasets that were used in this study. Section 5 describes the various building blocks of ACRE. Section 6 presents our experimental results on yield estimation and crop recommendation. Section 7 concludes the paper and provides several directions for future work.

2   Review of Relevant Work

In this section, we provide a review of relevant literature. Shi, Wang, and Fang [28] have selected 1176 papers in the area of *Artificial Intelligence for Social Good* from leading publication venues during the years 2008–2019 and provide a summary of the research. One of the key application areas explored in this paper is digital agriculture. Many papers surveyed here address the problems of yield prediction and crop recommendation. A significant amount of research has been reported on the effects of soil characteristics, climatic conditions, and geography on agricultural productivity.

2.1   Relevant Work in Crop Recommendation Systems

Von Lucken and Brunelli [12] use multi-objective evolutionary algorithms to select the best crop to plant for sustainable land use based on the soil data. The costs of fertilizing and liming, cultivation, and the expected fluctuation of total return are among the optimization criteria. This work provides a method based on multi-objective evolutionary algorithms to help select an optimal cultivation plan by taking into account five crop options and five objectives. This work considers a multi-objective crop selection problem having the following objectives: to minimize the cost of fertilization and lime application, minimize the total cost of production, maximize the average return, maximize the worst-case return, and minimize the standard deviation of returns.

Priyadharshini et al. [15] has proposed a system to assist the farmers in crop selection by considering factors such as sowing season, soil, and geographical location. To recommend a suitable crop to the user, the proposed system considers environmental data such as rainfall, temperature, and geographical location in terms of the characteristics such as soil type, pH value, and nutrient concentration. The paper seeks to develop a robust model that can accurately estimate crop sustainability in a given state for a given soil type and meteorological circumstances, and make recommendations for the best crops in the area to benefit the farmer. They use the following datasets: (1) Yield Dataset: contains yield for 16 major crops grown across all the states in kg per hectare; (2) Cost of Cultivation Dataset: provides the cost of cultivation for each crop in Rupees per hectare; (3) Modal price of crops: gives the average market prices for these crops over a period of two months; (4) Price of Crops: gives the current market price of the crops in Rs per hectare; (5) Soil Nutrient Content Dataset: contains Nitrogen content, Phosphorous content, Potassium content, and average pH value; and (6) Rainfall Temperature Dataset: contains crops, max, and min rainfall, max and min temperature, and pH values. The proposed system is implemented with linear regression and neural network and compared with K-Nearest Neighbours (KNN), KNN with cross validation, Decision Tree, and Naive Bayes.

Pudumalar et al. [16] has proposed a recommendation system through an ensemble model with majority voting technique using different machine learning tools to recommend a crop for the site specific parameters with high accuracy and efficiency. Their dataset contains the soil specific attributes collected for Madurai district tested at soil testing lab, Madurai, Tamilnadu, India. The crops considered in this paper include millets, groundnut, pulses, cotton, vegetables, banana, paddy, sorghum, sugarcane, coriander. Different machine learning models give the best suitable crop, based on the given input, and then the crop which is selected by most of the models is recommended to the farmer which is simply a majority voting rule.

Madhuri and Indiramma [13] introduce a recommendation system which uses an artificial neural networks (ANN) with four layers for recommending the best suitable crop. The crops considered are maize, finger-millet, rice, and sugarcane. In this work, they have considered the data for two locations namely Hadonahalli and Durgenahalli of Doddaballapur, Karnataka, India. They have the predefined conditions which are favourable for these crops and train their neural network using this data.

We have only presented representative papers on crop recommendation. We refer the reader to papers cited in the above papers and in the paper by Shi, Wang, and Fang [28] to probe further.

2.2   Relevant Work in Crop Yield Prediction

There is abundant literature on the crop yield prediction problem and we point the reader to a comprehensive survey on crop yield prediction [24].

The paper by Sharma, Rai, and Krishnan [19] presents a method to predict crop yields from publicly available satellite imagery. The idea is to learn a deep neural network model for predicting the wheat crop yield for tehsils (also called taluk - a local unit of administration covering a town and villages within a radius of 20 to 30 km) in India. The method uses raw satellite imagery and notably, there is no need to extract any hand-crafted features or perform dimensionality reduction on the images. A byproduct of this work is to create a new dataset comprising a sequence of satellite images and the exact crop yield for the years 2001–2011 covering a total of 948 tehsils. This dataset is used to train and evaluate the proposed approach on tehsil level wheat predictions. The model outperforms existing methods by over 50%. Additional contextual information such as the location of farmlands, water bodies, and urban areas improves the yield estimates.

You et al. [27] develops a scalable, accurate, and low-cost technique for predicting agricultural yields using publicly available remote sensing data. They suggested a strategy based on deep learning ideas rather than the hand-crafted features commonly utilized in the remote sensing sector. They also present a dimensionality reduction strategy that enables them to train a convolutional neural network or a long-short term memory (LSTM) network and learn valuable features automatically even when labeled training data is scarce. They include a Gaussian Process component to model the data's spatio-temporal structure and increased accuracy explicitly.

The paper by Shahhosseini et al. [18] shows improvements in yield prediction accuracy when ML models are used in conjunction with additional side information obtained from a simulation cropping systems model. Soil water related variables (particularly growing season average drought stress and average depth to water table) when supplied as input to the ML model lead to better predictions. The study is based on data from the central US Corn Belt. Khaki, Wang, and Archontolis [10] propose a deep learning framework for agricultural yield prediction based on environmental data. They use convolutional neural networks (CNNs) and recurrent neural networks (RNNs). Using historical data, the proposed CNN-RNN model is used in conjunction with other popular methods such as random forest (RF), deep fully connected neural networks (DFNN), and LASSO (Least Absolute Shrinkage and Loss Operator) to forecast corn and soybean yield across the entire Corn Belt (including 13 states) in the United States for the years 2016, 2017, and 2018. The CNN-RNN model is created to capture the temporal dependence of environmental factors and the genetic improvement of seeds over time. Their model is a combination of CNNs, fully connected layers, and RNNs. They have predicted average yield for corn and soybean by using the RNN layer which has a time length of 5 years since they consider 5-year yield dependencies.

An optimization model created by Awad [1] is able to estimate crop yield by increasing remote sensing data availability. The optimization model is created by using a well-known algorithm, Trust-Region Methods for Nonlinear Minimization, that fits available data to an exponential equation. The experimental results prove the accuracy and reliability of the new model in estimating the crop yield in the case of missing remote sensing data. Using potato as the crop, the paper shows that the lack of remote sensing data is not a serious handicap for yield prediction.

Fan et al. [3] has introduced a novel graph-based recurrent neural network (GNN) for crop yield prediction, which incorporates both geographical and temporal structure. Their method is trained, validated, and tested on over 2000 counties from 41 states in the United States of America, covering years from 1981 to 2019. The paper compares 11 representative machine learning models, including GNN and GNN-RNN, on US county-level crop yields for corn and soybean. They are able to achieve an impressive improvement over existing models. The paper clearly brings out the importance of exploiting geospatial context in making yield predictions.

A pilot study by the International Crops Research Institute for the Semi-Arid Tropics [6] in India instructed farmers to delay planting by three weeks using predictive models based on climate and weather data and increased the yield by 30 percent. In the Indian state of Andhra Pradesh, they tested a new sowing application for farmers combined with a personalized village advisory dashboard, with the results showing a 30 percent greater average yield per hectare. During the pilot, farmers received 10 sowing advisories containing important information such as sowing recommendations, seed treatment, optimum sowing depth, preventive weed management, land preparation, farmyard manure application, harvesting recommendations, shade drying of harvested pods, and storage.

2.3 Positioning of Our Work

In most research efforts in the current literature, usually only certain subsets of crops have been taken into account while making recommendations. Furthermore, the recommendation has been for a single crop to the farmer based on some objective function. The papers do not consider inter-cropping and do not recommend portfolios of multiple crops after assessing risk to the farmer. ACRE computes the expected utility to the farmer taking into account profit and risk. ACRE also provides several options for utility functions. Most importantly, ACRE uses ideas from the financial investments literature to generate crop portfolio recommendations. In particular, ACRE uses Sharpe ratio [20,21], a widely used financial portfolio selection metric, to obtain a ranking of crop portfolios. Section 3 presents an overview of the Sharpe ratio.

In terms of yield prediction, we have seen that many methods have been employed in the literature. We have used ensemble methods that combine different machine learning and deep learning models to produce accurate yield estimates.

3 Sharpe Ratio

The *reward-to-variability ratio*, popularly known as the Sharpe Ratio [20,21] is a popular investment portfolio performance measure. It compares the return of an investment with its risk. This is done by comparing the performance of the investment, such as a portfolio, to a risk-free asset after adjusting for its risk. It is also used in other similar contexts where one has to choose between a number of options that may yield different returns. In this paper, we have used the Sharpe ratio to quantify the risk-reward of potential crop-portfolios that may be suggested to the farmer. The investment portfolio in the agricultural context is the portfolio of crops grown by the farmer. The return from the portfolio is the revenue from sale of the harvest. The risk-free asset is the cost saved by the farmer by not growing anything at all. It can additionally include a fixed risk free rent begotten by leasing out the land. The risk here involves the risk in all of the yield, the duration go growing period and the price of the crops in the portfolio, which translates to a direct uncertainty in profit. The Sharpe Ratio is a single representative number that is drawn from the mean and variance of the differential with a standard benchmark portfolio, making it a very useful tool in portfolio recommendation. When we choose the portfolio with the highest Sharpe Ratio we ensure that additional risk is compensated with additional reward. This ensures that an optimal portfolio of crops that maximises profit while accounting for risk is found.

The Sharpe Ratio can be computed Ex-Post (that is, as late as possible when most information required is available) or as Ex-Ante (that is, as early as possible, working with probabilistic estimate of required information). In ACRE, we use the historical values of various input parameters to predict the future yield and price. These are used to calculate the Ex-Ante Sharpe Ratio, which helps us make recommendations to the farmer depending on the farmer's risk preferences. The Ex-Ante Sharpe Ratio is calculated as the ratio of the mean and the standard deviation of the differential:

$$S \equiv \frac{\bar{d}}{\sigma_d}$$

where d is the differential between the returns generated by the recommended crop portfolio, P and the benchmark (e.g., minimum rental income), B. If the returns generated by portfolio P and the benchmark are represented by R_P and R_B respectively, then the differential is calculated as

$$\tilde{d} \equiv \tilde{R}_P - \tilde{R}_B$$

We use tildes in the equation above to signify that the exact values of the returns, R_P and R_B might not be known before hand (Ex-Ante). The Sharpe Ratio can also be defined as the ratio of expected added return per unit of added risk to the benchmark.

In the absence of a benchmark, rather than taking a differential return ratio, one may also take a direct ratio of the return's mean and standard deviation. We have used this approach to compare the various crop portfolios.

4   Data Collection and Curation

We have used a variety of datasets in this project. The datasets cover yield, weather, and soil characteristics. All these data are drawn from Indian Government websites. The yield data is from [2,7]; the price data is from [2]; and the weather data is from [4].

4.1   Yield Data

We are using the yield data provided by VDSA and ICRISAT. This data is available for various crops from 1966 to 2011. We worked on Uttar Pradesh yield data since that was more complete when compared to the other states. The granularity of the data is district-wise. The agmarknet website provides the arrival data, which has the production quantity of different crops in the mandis. The granularity of this data is mandi-wise in every district. The arrival data for different products unfortunately does not contain the area (in hectares) from which the products have arrived.

We have also used data for yield available at [7], and its granularity is district-wise. It includes state, district, area in hectares, and production in tonnes. It contains the average yield of a crop for a particular year. However, this data suffers from many missing values. We have used standard techniques in data imputation [8,26] to fill the missing values.

4.2 Weather Data

We have used the weather data from [4]. This data includes environmental parameters such as temperature, rainfall, humidity, and sunlight. This weather data is available for every 0.25-degree change in latitude and longitude. We have taken the weather data, month-wise, for Kharif and Rabi seasons, separately.

4.3 Soil Data

We have taken the data available at [4] for the type of soil. This parameter is the texture (or classification) of soil used by the land surface scheme of the ECMWF (European Centre for Medium-Range Weather Forecasts) Integrated Forecasting System (IFS) to predict the water holding capacity of soil in soil moisture and runoff calculations. It is derived from the root zone data (30–100 cm below the surface) of the FAO/UNESCO Digital Soil Map of the World, DSMW (FAO, 2003), which exists at a resolution of $5' \times 5'$ (about 10 km). The seven soil types are 1: Coarse, 2: Medium, 3: Medium fine, 4: Fine, 5: Very fine, 6: Organic, and 7: Tropical organic. A value of 0 indicates a non-land point. This parameter does not vary in time. This data [4] is available for every 0.25-degree change in latitude and longitude. The Agro-Ecological Sub-Regions (AESR) index represents which type of land and which type of soil are present in a particular region based on latitude and longitude.

5   Building Blocks of ACRE

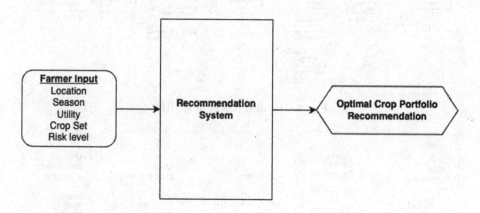

Fig. 1. Farmer's view of ACRE

The basic idea of ACRE is to make crop portfolio recommendations to farmers. Towards this end, we collect various input data from the farmer, which could be based on their traditional or cultural practices.

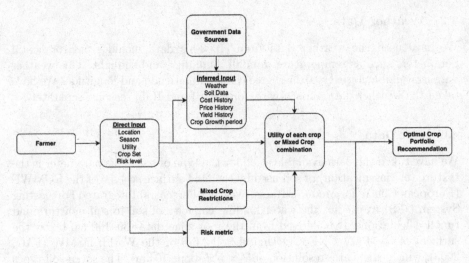

Fig. 2. Data flow model

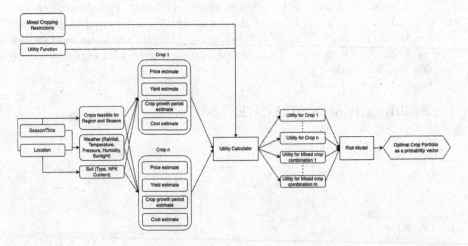

Fig. 3. Detailed model of ACRE

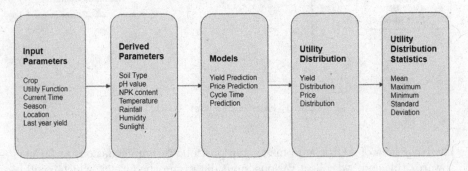

Fig. 4. Architecture of the utility calculator

5.1   Input Parameters

ACRE uses many input parameters. These could be classified into direct parameters and inferred parameters.

- **Direct Parameters:** The farmer can directly select these parameters. These include state and district, subset of crops, season (Kharif, Rabi, Whole Year), utility functions, irrigation facilities, crop rotation cycle, area (in hectares), human labour cost, etc. Each of these would have default options and will apply if the farmer is unable to select some of the parameters. In fact, some of these could even be inferred if the farmer is unable to select them.
- **Inferred Parameters:** These parameters are derived from direct parameters. These include temperature, rainfall, sunlight, humidity, soil type and nutrients, and crop price. ACRE infers these parameters. Some of the direct parameters that are not selected by the farmer are also inferred by the system.

ACRE takes all available inputs and builds models to predict yield, price, cost, and duration of crop growing period. The farmer can choose the type of utility (yield, profit, or risk). The models are used to compute the utility distribution (triangular distribution) based on the utility function chosen by the farmer. Based on the utilities, ACRE recommends a possible portfolio of crops based on the Sharpe ratio considering expected return and the risk associated with it. Figure 1 shows the farmer's viewpoint of ACRE. The farmer sees ACRE as a black box, which takes various inputs from the farmer and outputs a crop portfolio.

5.2   Utility Calculator

Figure 4 shows the basic architecture of the utility calculator. The utility calculator computes the utility distribution for the farmer based on the parameters provided and the utility function chosen by the farmer. The utility calculator outputs the expectation and standard deviation of the utilities, which are used in recommending a suitable portfolio of crops to the farmer.

Utility Functions. The farmer can choose from the following utility functions.

Profit (Considering Growing Period)

$$\text{Utility} = \frac{Y * P - C}{T} \tag{1}$$

Here, Y denotes yield in tonnes per hectare, P denotes crop price in Rupees per tonne, C denotes the cost of cultivation per hectare, and T denotes duration of the growing period for a crop in months. This utility function computes the profit for the farmer on a particular crop. It helps maximize the farmer's profit by considering the crop logistics cost and growing period on a monthly basis. This utility function assumes that farmers can grow any crop in any combination. We can also consider the variance in the yield and price while generating the profit utility distribution for various crops.

Profit (Without Considering Growing Period)

$$\text{Utility} = Y * P - C \qquad (2)$$

This utility function does not consider the duration of the growing period, because, sometimes a farmer could grow only one crop in the entire year due to specific constraints like scarcity of resources, local policies, etc. In this case the actual duration of crop growing period would not matter as long as it is less than one year. One may also account for rate of interest to be more accurate.

Yield. This utility function simply considers the expected yield per hectare. This can be used for scenarios where only the yield of a particular crop is the determining factor. The farmer may not be interested in the profit from these crops and might be growing them for reasons such as personal consumption, adding nutrients to the soil, cattle feed, etc.

Crop Portfolio Recommendation. Based on the utility distribution and risk profile of the farmer, ACRE recommends a convex combination of crops to the farmer.[1] Cultural and traditional practices of the farmer can be taken into account in this portfolio of crops by eliciting from the farmer which crops the farmer is interested in growing. Also, we could take last year's yields for the farmer's crops and check whether or not the recommendation given by ACRE is aligned with the farmer's traditional practices or cultural preferences. Further, instead of providing a single recommendation, ACRE can provide a ranked list of multiple best portfolio recommendations and the farmer could choose based on any personal or cultural preferences.

Figure 2 contains a depiction of the data flow in ACRE. The direct inputs such as location of the farm, set of crops under consideration, etc., could be provided by the farmer (with expert assistance, if required). This is used to download the required inferred inputs such as local weather, price history in nearby mandis, etc. There are some crops that cannot be grown together and the mixed crop restrictions take such constraints into account. These constraints, the direct inputs, and the indirect inputs are used to calculate the utility to the farmer provided by various crop portfolios. Depending on the farmer's risk preferences, one or more of these crop portfolios are recommended to the farmer as the final output.

The various building blocks of the recommendation system are shown in Fig. 3. The utility calculator computes the utility distribution, from the yield and price predictions, for different feasible combinations of the crops provided as input. The risk model takes into account the utility distributions for all these specified crops and the risk profile of the farmer, and, recommends a suitable crop or combination of crops. The crop portfolio is a convex combination that specifies the fractions in which to grow the different crops.

---

[1] A convex combination of numbers is a weighted sum of the numbers with weights in the range $[0, 1]$ and the weights summing to 1.

6 Experiments and Results

In the experiments reported here, we have worked with profit maximization utility for farmers. To obtain the utility distribution of a particular crop, we need yield, price, cost, and growing period of a crop. Currently, we have considered the yield distribution and price distribution of a crop. Duration of growing period and costing models have not been considered in this work, but can be looked into for more accurate predictions. We have considered the price of a crop during the harvesting period of the particular crop. We have obtained the price distribution using the price variations in these months. For getting the yield distribution of a crop, we have used the ensemble technique in which we have trained multiple machine learning models and considered a 10% deviation from the predicted yield to get maximum and minimum values (based on our reading of historical data).

6.1 Crop Yield Prediction

For our pilot experiments, we have considered three crops - rice, maize, and groundnut - for the Kharif season and three crops - wheat, barley, and masur-dal - for the Rabi season. We have experimented with several standard machine learning and deep learning models [11] to predict yield. These include: Poly-nomial Regression, Random Forest Regression, Deep Neural Network Regres-sion (DNN), Convolutional Neural Network on Deep Neural Network Regression (CNN-DNN), Long Short Term Memory Regression (LSTM), and Convolutional Neural Network on Long Short Term Memory Regression (CNN-LSTM) where a convolutional neural network is used for extracting the embeddings of the fea-tures. The features used for predicting the yield include latitude, longitude, land characteristics such as the AESR index, and weather parameters such as rainfall, temperature, sunlight, and humidity.

The results are measured and compared using different metrics. These met-rics include Root Mean Square Error (RMSE), 95% Confidence Interval, 90% Confidence Interval, R^2 Score, Pearson Correlation Coefficient, and Coefficient of Variation. The confidence interval represents the percentage of the predicted values within this range of actual test values. The correlation coefficient repre-sents how much the predicted and actual test values are related. The coefficient of Variation is the ratio of the standard deviation of predicted values and the expectation of actual test values.

For training the regression models, we have taken the yield data for Uttar Pradesh from 1982 to 2008, and for testing we have taken the data from 2009 to 2011. We have developed all the regression models mentioned above for all the crops separately. Wheat has the highest stability in crop yield over the years and the regression models expectedly predicted its yield very accurately.

Table 1. Yield prediction results for Wheat for various regression models.

Regression models	RMSE	95% confidence interval	90% confidence interval	R2 score	Correlation coefficient	Coefficient of variation
Polynomial regression	0.33	35.53	63.16	0.76	0.89	0.14
Random forest	0.22	45.39	76.64	0.89	0.94	0.09
DNN	0.25	46.38	74.34	0.86	0.93	0.10
CNN-DNN	0.27	44.41	71.38	0.84	0.92	0.11
LSTM	0.21	44.69	72.67	0.89	0.95	0.09
CNN-LSTM	0.22	47.91	74.60	0.88	0.94	0.09

Table 1 provides the metrics for all the regression models for wheat. The random forest, DNN, and CNN-LSTM regression models perform better than the other regression models. The above models have been explored for other crops as well. The random forest Regression Model is found to outperform other regression models for almost all the crops. Table 2 provides the results for all crops using the random forest model. The estimates for groundnut and maize are not as good as other crops due to the inherent variability in the yields of these two crops.

Table 2. Yield prediction results for all crops using the random forest model.

Crops	RMSE	95% confidence interval	90% confidence interval	R2 score	Correlation coefficient	Coefficient of variation
Barley	0.28	31.35	54.13	0.81	0.90	0.16
Wheat	0.22	45.39	76.64	0.89	0.94	0.09
Masur-Dal	0.14	33.93	57.14	0.49	0.71	0.17
Rice	0.23	40.79	66.78	0.81	0.90	0.12
Maize	0.29	18.92	39.53	0.55	0.74	0.24
Groundnut	0.29	20.65	33.77	0.48	0.70	0.44

Results Using Ensembling: Ensemble techniques integrate multiple base models to create a single best-fit predictive model. Ensemble methods, if properly deployed, provide higher predictive accuracy compared to any individual model. We have experimented with the ensemble technique in which we have taken all the combinations of random forest, DNN, CNN-DNN, and LSTM regression models for ensembling. For almost all the crops, ensemble techniques are found to perform better than the individual models. The ensemble technique which consists of random forest and DNN is found to perform better than other ensemble techniques for almost all the crops. Table 3 shows the different ensembles of several regression models and their performance for yield prediction of wheat. The ensembles (RF, DNN), (RF, CNN-DNN), and (RF, LSTM) perform better than the others. Table 4 shows the results of the ensemble method consisting of RF and DNN for all the other crops.

Table 3. Yield prediction results for wheat: ensemble technique

Models	RMSE	95% confidence interval	90% confidence interval	R2 score	Correlation coefficient	Coefficient of variation
RF, DNN	0.13	66.15	89.74	0.96	0.98	0.06
RF, CNN-DNN	0.14	61.03	84.87	0.95	0.98	0.06
RF, LSTM	0.12	64.10	89.74	0.96	0.98	0.06
DNN, CNN-DNN	0.18	53.33	76.41	0.92	0.96	0.08
DNN, LSTM	0.15	53.85	82.56	0.94	0.97	0.07
CNN-DNN, LSTM	0.16	51.79	78.97	0.93	0.97	0.08
RF, DNN, CNN-DNN	0.14	59.74	84.87	0.95	0.97	0.07
RF, DNN, LSTM	0.13	62.56	88.21	0.96	0.98	0.06
RF, CNN-DNN, LSTM	0.13	61.28	85.38	0.95	0.98	0.06
DNN, CNN-DNN, LSTM	0.16	53.85	80.00	0.94	0.97	0.07
RF,DNN,CNN-DNN,LSTM	0.14	59.74	86.41	0.95	0.98	0.06

6.2   Results on Profit Utilities

As mentioned in Sect. 6.1, we have chosen three possible crops each for the Rabi and Kharif seasons. We have computed the profit utility distributions for the farmer consisting of average profit, maximum profit, minimum profit, and actual profit of the crops. We have also computed the coefficient of risk in profit (standard deviation in profit divided by the expected profit value for the crops).

Table 5 shows the profit utility distributions for the Kharif and Rabi seasons for the year 2011. The maximum profit and minimum profit are computed using the 10% deviation in predicted yield. We have chosen 10% based on our observation of ground truth data. Note that different crops have different durations of growing periods and for normalisation, the profit is computed in rupees per hectare per month. In the Kharif season, groundnut has the maximum profit but also shows the highest variation in profit, representing the highest risk. In contrast, maize has the lowest maximum profit but the lowest risk. Similarly, in the Rabi season, masur dal shows the highest maximum profit and also the lowest risk and turns out to be the best crop for Rabi season.

Table 6 shows the profit utility distributions for the Kharif and Rabi seasons for the year 2010. Here, again, groundnut has the highest maximum profit as well as the highest risk. Maize has the lowest maximum profit but it is the most stable crop in the Kharif season. In the Rabi season, masur-dal is again showing the highest maximum profit with the lowest risk in comparison to all the other crops in the Rabi season.

Table 4. Yield prediction results for the ensemble technique consisting of RF and DNN for all the crops.

Crops	RMSE	95% confidence interval	90% confidence interval	R2 score	Correlation coefficient	Coefficient of variation
Barley	0.14	43.55	72.93	0.90	0.94	0.09
Wheat	0.13	66.15	89.74	0.96	0.98	0.06
Masur-Dal	0.14	32.73	53.57	0.43	0.66	0.17
Rice	0.13	52.82	84.35	0.92	0.96	0.07
Maize	0.14	33.34	62.69	0.78	0.90	0.13
Groundnut	0.28	12.45	30.81	0.49	0.70	0.43

Table 5. Profit utility distribution for different crops for the year 2011.

Crop	Season	Predicted profit	Maximum profit	Minimum profit	Actual profit	Variance (risk)
Rice	Kharif	5717.89	6444.15	1592.598	6172.70	0.19
Maize	Kharif	5875.08	6312.06	5456.69	6423.62	**0.03**
Groundnut	Kharif	5444.50	**21104.88**	1431.89	5844.84	0.78
Barley	Rabi	4632.27	9139.18	3939.41	4579.91	0.25
Wheat	Rabi	6571.12	7430.05	5577.08	6743.59	0.06
Masur-Dal	Rabi	16456.15	**17218.37**	15492.85	18889.67	**0.02**

Table 6. Profit utility distribution for different crops for the year 2010.

Crop	Season	Predicted profit	Maximum profit	Minimum profit	Actual profit	Variance (risk)
Rice	Kharif	5434.73	6118.97	4842.20	6402.49	0.05
Maize	Kharif	5449.76	5880.31	5036.29	6930.78	**0.03**
Groundnut	Kharif	4507.94	**20187.35**	494.427	4884.12	0.94
Barley	Rabi	3945.59	4449.53	3171.27	3876.45	0.07
Wheat	Rabi	6281.94	7427.60	5473.42	7030.19	0.06
Masur-Dal	Rabi	8067.66	**8668.42**	7366.42	19000.67	**0.03**

Table 7. Individual crop recommendation based on maximum profit and minimum risk.

Year	Season	Maximizing profit	Minimizing risk
2011	Kharif	Groundnut	Maize
2011	Rabi	Masur-Dal	Masur Dal
2010	Kharif	Groundnut	Maize
2010	Rabi	Masur-Dal	Masur-Dal

6.3 Recommendation of Individual Crops

Based on the profit utility distribution, the variance or risk associated with the profit, and the farmer's risk profile, ACRE recommends the best possible crop to the farmer for the particular season. Some farmers are risk-loving; they would prefer the crop which maximizes their profit irrespective of the risk associated with that crop. ACRE would typically suggest such farmers to grow groundnut in the Kharif season. On the other hand, there are also farmers who are very risk-averse; these farmers prefer stable, even if somewhat low, returns from the crop. A good suggestion for such farmers in the Kharif season would be to grow maize. For the Rabi season, masur-dal would be best suited to both these types of farmers. Table 7 shows single crop per season recommendations for the years 2010 and 2011 for Kharif and Rabi seasons. As the risk profile varies between these two extremes, the suggestions made by ACRE vary over different fractions of the crops to be sown.

6.4 Sharpe Ratio Based Crop Portfolio Recommendation

To make recommendations to farmers whose risk profile lies somewhere between the two extremes (risk loving and risk averse), ACRE analyzes the results of sowing different convex combinations of crops. The Sharpe ratio is used to find the most suitable combination or portfolio of crops for the farmers for a particular season.

Table 8. Sharpe ratio for different crop portfolios for Kharif season for the years 2009, 2010, and 2011.

Portfolios (Rice, Maize, Groundnut)	Sharpe ratio 2009	Sharpe ratio 2010	Sharpe ratio 2011
(0.0, 0.0, 1.0)	0	0	0
(0.0, 0.2, 0.8)	0.84	0.89	1.07
(0.0, 0.4, 0.6)	2.23	2.37	2.85
(0.0, 0.6, 0.4)	5	5.33	6.4
(0.0, 0.8, 0.2)	13.13	14.03	16.83
(0.0, 1.0, 0.0)	73.42	84.91	100
(0.2, 0.0, 0.8)	1.15	0.89	1.03
(0.2, 0.2, 0.6)	2.64	2.37	2.79
(0.2, 0.4, 0.4)	5.63	5.33	6.27
(0.2, 0.6, 0.2)	14.44	14.09	16.22
(0.2, 0.8, 0.0)	93.29	99.71	66.77
(0.4, 0.0, 0.6)	3.05	2.36	2.68
(0.4, 0.2, 0.4)	6.24	5.32	5.99
(0.4, 0.4, 0.2)	15.63	14.05	14.61
(0.4, 0.6, 0.0)	100	100	36.86
(0.6, 0.0, 0.4)	6.83	5.29	5.59
(0.6, 0.2, 0.2)	16.69	13.93	12.57
(0.6, 0.4, 0.0)	90.7	85.42	23.74
(0.8, 0.0, 0.2)	17.61	13.72	10.56
(0.8, 0.2, 0.0)	77	68.47	16.81
(1.0, 0.0, 0.0)	65.32	54.93	12.57
(0.33, 0.33, 0.33)	7.9	7.09	8.06
(0, 0.5, 0.5)	3.34	3.56	4.27
(0.5, 0, 0.5)	4.57	3.54	3.92
(0.5, 0.5, 0)	96.66	93.76	29.13

In ACRE, we use the Sharpe ratio on profit maximization and it is computed by dividing the expected profit value by the standard deviation in the profit. For easier comparison, the values of the Sharpe ratio for different portfolios have been scaled from 0 to 100, where 0 is the lowest and 100 is the highest Sharpe ratio for a crop profile. There are four possible extreme categories for a portfolio: (a) the highest return and least risk (b) lowest return and least risk (c) highest return and highest risk (d) lowest return and highest risk. The categories (a) and (b) would have high Sharpe ratio values, while (c) and (d) would have low Sharpe ratio values. Clearly, the categories (a) and (b) are favourable for small and marginal farmers.

Table 9. Sharpe ratio for different crop portfolios for Rabi season for the years 2009, 2010, and 2011.

Portfolios (Barley, Wheat, Masur-Dal)	Sharpe ratio 2009	Sharpe ratio 2010	Sharpe ratio 2011
(0.0, 0.0, 1.0)	87.31	69.7	93.62
(0.0, 0.2, 0.8)	99.79	85.99	100
(0.0, 0.4, 0.6)	100	79.6	96.69
(0.0, 0.6, 0.4)	84.22	52.75	77.65
(0.0, 0.8, 0.2)	63.96	24.64	51.58
(0.0, 1.0, 0.0)	47.19	3.01	29.32
(0.2, 0.0, 0.8)	56.3	82.06	76.12
(0.2, 0.2, 0.6)	56.96	100	73.83
(0.2, 0.4, 0.4)	53.14	80.63	63.07
(0.2, 0.6, 0.2)	45.64	43.2	45.31
(0.2, 0.8, 0.0)	36.86	13.12	26.87
(0.4, 0.0, 0.6)	26	84.52	41.96
(0.4, 0.2, 0.4)	24.38	94.23	35.07
(0.4, 0.4, 0.2)	22	62.16	26
(0.4, 0.6, 0.0)	19.04	24.4	15.95
(0.6, 0.0, 0.4)	12	65.51	20.32
(0.6, 0.2, 0.2)	10.63	60.7	14.38
(0.6, 0.4, 0.0)	9.06	30.44	7.95
(0.8, 0.0, 0.2)	4.55	31.77	7.8
(0.8, 0.2, 0.0)	3.46	20.93	3.09
(1.0, 0.0, 0.0)	0	0	0
(0.33, 0.33, 0.33)	30.32	83.56	39.28
(0, 0.5, 0.5)	93.54	67.43	88.94
(0.5, 0, 0.5)	17.78	77.88	29.61
(0.5, 0.5, 0)	13.28	28.79	11.46

Table 8 shows the Sharpe ratios for different portfolios of crops for the Kharif season for the years 2009, 2010, and 2011. The highest Sharpe ratio is calculated in the years 2009 and 2010 for the portfolio of rice, maize, and groundnut with the convex combination of (0.4, 0.6, 0.0) and the convex combination (0.0, 1.0, 0.0) yields the highest Sharpe ratio for the year 2011. Thus, in 2011, growing only maize yields the highest Sharpe ratio.

Table 9 shows the Sharpe ratios for different portfolios of crops for the Rabi season for the years 2009, 2010, and 2011. In the years 2009, 2010, and 2011 the highest Sharpe ratio for Rabi season is achieved respectively by the convex combinations (0.0, 0.4, 0.6), (0.2, 0.2, 0.6), and (0.0, 0.2, 0.8). It is evident that the optimal portfolios are a mixture of more than one crop. Instead of growing the crop which gives highest expected profit, it is better to take an optimal portfolio to reduce risk.

6.5   Socio-Cultural Factors in Crop Recommendation

Thus far, only factors such as weather and soil conditions have been considered in ACRE. There could be other factors such as water, availability of labour and availability of farm equipment that could be taken into account, subject to data availability. There are, however, certain socio-cultural factors where it is hard to enumerate, or even find, the data. Some of these factors include:

- Do the Government policies favour the recommended crops?
- Are there any incentives available for the recommended crops?
- Do the traditional beliefs and family conditions (and even superstitions) favour the recommended crops?
- What do the family members, friends, neighbours, and relatives think about the recommended crops?
- What were the previous experiences of the farmer with the recommended crops?

The best way to handle such socio-cultural factors would be to get a ranked list of multiple recommendations from ACRE and provide assistance to the farmer in choosing the best recommendation while keeping these factors in mind.

7   Summary and Future Work

ACRE recommends a near optimal portfolio of crops by computing various utility values for farmers using data on crop characteristics such as yield, price, costing, and crop growing period. The recommendation system computes the estimated profit of each crop and recommends a portfolio that maximises profit while accounting for the risk in yield, price, and duration of growing period of the crop.

This project is an initial effort in crop recommendation. There are many ways ACRE could be enhanced. This involves creating more effective mathematical models or distributions to represent yield, price, and cost data accurately. The accuracy can be further increased with better forecasting models. Apt utility functions can be designed, which better cover all of the farmers' requirements. We have currently considered only soil and weather factors. There are many other factors that need to be taken into account including demand side factors and socio-cultural factors [22]. We have used the Sharpe ratio for portfolio recommendation; there are many refinements of Sharpe ratio [17,23] that could potentially give better results. Furthermore, risk in water supply, storage, weather parameters and procuring agricultural inputs like seeds, fertilizers and pesticides have not been considered. In portfolio recommendation we have not taken into account an important factor, namely, mixed cropping constraints. Our experiments are rather small-scale with only three candidate crops each for Rabi and Kharif seasons. Also, it would make sense to recommend a portfolio of crops for the entire year rather than for a season. We have not used a price prediction model; it would be useful to integrate a long horizon price prediction with ACRE to enhance the quality of recommendation. Above all, there is a need to interact with a wide cross section of farmers, especially small and marginal farmers, to understand their constraints and expectations from a tool like ACRE.

The next crucial step in implementing ACRE on the ground is to develop a user-friendly mobile app that obtains the needed inputs from the farmer. We are currently developing the same even as the computation and inference capabilities of ACRE get enhanced.

Acknowledgment. We gratefully acknowledge the support provided by the National Bank for Agriculture and Rural Development (NABARD), Government of India, through a research grant for carrying out this work. We must thank the discussions with Prof. Lalith Achot and Prof Vedamurthy in getting several of our questions clarified in crop recommendation. The third author (Mayank Ratan Bhardwaj) wishes to thank the Ministry of Education, Government of India, for providing the Doctoral Fellowship.

References

1. Awad, M.M.: Toward precision in crop yield estimation using remote sensing and optimization techniques. Agriculture **9**(3), 54 (2019)
2. Ministry of Agriculture Directorate of Marketing & Inspection (DMI) and Government of India Farmers Welfare. Agmarknet. https://agmarknet.gov.in/
3. Fan, J., Bai, J., Li, Z., Ortiz-Bobea, A., Gomes, C.P.: A GNN-RNN approach for harnessing geospatial and temporal information: application to crop yield prediction. In: Proceedings of the AAAI Conference on Artificial Intelligence, vol. 36, pp. 11873–11881 (2022)
4. European Centre for Medium-Range Weather Forecasts. The climate data store. https://climate.copernicus.eu/climate-data-store
5. India Brand Equity Foundation. Agriculture in india: Information about Indian agriculture & its importance. https://www.ibef.org/industry/agriculture-india.aspx
6. ICRISAT. Microsoft and ICRISAT's intelligent cloud pilot for agriculture in Andhra Pradesh increase crop yield for farmers. https://www.icrisat.org/microsoft-and-icrisats-intelligent-cloud-pilot-for-agriculture-in-andhra-pradesh-increase-crop-yield-for-farmers/
7. Digital India Initiative. Agriculture. https://data.gov.in/sector/Agriculture
8. Jäger, S., Allhorn, A., Biebmann, F.: A benchmark for data imputation methods. Front. Big Data (2021)
9. Khaki, S., Wang, L.: Crop yield prediction using deep neural networks. Front. Plant Sci. **10**, 621 (2019)
10. Khaki, S., Wang, L., Archontoulis, S.V.: A CNN-RNN framework for crop yield prediction. Front. Plant Sci. **10**, 1750 (2020)
11. LeCun, Y., Bengio, Y., Hinton, G.: Deep learning. Nature **521**(7553), 436–444 (2015)

12. Von Lücken, C., Brunelli, R.: Crops selection for optimal soil planning using multi-objective evolutionary algorithms. In: AAAI-2008, 22nd International Conference of the American Association for Artificial Intelligence, pp. 1751–1756 (2008)

13. Madhuri, J., Indiramma, M.: Artificial neural networks based integrated crop recommendation system using soil and climatic parameters. Indian J. Sci. Technol. **14**(19), 1587–1597 (2021)

14. National Portal of India. Agriculture. https://www.india.gov.in/topics/agriculture

15. Priyadharshini, A., Chakraborty, S., Kumar, A., Pooniwala, O.R.: Intelligent crop recommendation system using machine learning. In: Proceedings of the Fifth International Conference on Computing Methodologies and Communication (ICCMC 2021) IEEE Xplore Part Number: CFP21K25-ART (2021)

16. Pudumalar, S., Ramanujam, E., Harine Rajashree, R., Kavya, C., Kiruthika, T., Nisha, J.: Crop recommendation system for precision agriculture. In: 2016 Eighth International Conference on Advanced Computing (ICoAC), pp. 32–36. IEEE (2017)

17. Scholz, H.: Refinements to the sharpe ratio: Comparing alternatives for bear markets. J. Asset Manag. **7**(5), 347–357 (1966)

18. Shahhosseini, M., Hu, G., Huber, I., Archontoulis, S.V.: Coupling machine learning and crop modeling improves crop yield prediction in the us corn belt. Sci. Rep. **11**(1), 1–15 (2021)

19. Sharma, S., Rai, S., Krishnan, N.C.: Wheat crop yield prediction using deep LSTM model. Technical report (2020)

20. Sharpe, W.F.: Mutual fund performance. J. Bus. **39**(1), 119–138 (1966)

21. Sharpe, W.F.: The sharpe ratio. J. Portf. Manag. **21**(1), 49–58 (1994)

22. Shekara, P.C., et al.: Farmer's Handbook on Basic Agriculture. Desai Fruits & Vegetables Pvt. Ltd., Navsari (2016)

23. Sortino, F.A., Price, L.N.: Performance measurement in a downside risk framework. J. Invest. **3**, 50–58 (1994)

24. Van Klompenburg, T., Kassahun, A., Catal, C.: Crop yield prediction using machine learning: a systematic literature review. Comput. Electron. Agric. **177**, 105709 (2020)

25. Vikaspedia. Critical factors to be considered for selection of crops (2022). https://vikaspedia.in/agriculture/crop-production/critical-factors-to-be-considered-for-selection-of-crops

26. Woźnica, K., Biecek, P.: Does imputation matter? Benchmark for predictive models. arXiv preprint arXiv:2007.02837 (2020)

27. You, J., Li, X., Low, M., Lobell, D., Ermon, S.: Deep gaussian process for crop yield prediction based on remote sensing data. In: Thirty-First AAAI Conference on Artificial Intelligence (2017)

28. Fang, F., Shi, Z.R., Wang, C.: Artificial intelligence for social good: a survey. Technical report, Carnegie Mellon University (2020)

Analyze the Impact of Weather Parameters for Crop Yield Prediction Using Deep Learning

Pragneshkumar Patel[1]([✉]), Sanjay Chaudhary[1], and Hasit Parmar[2]

[1] Ahmedabad University, Gujarat, India
{pragnesh.p,sanjay.chaudhary}@ahduni.edu.in
[2] L. D. College of Engineering, Gujarat, India

Abstract. Accurate crop yield prediction is one of the most important aspects for the agricultural policy decision for the policy makers and the farmers. However, prediction of crop yield depends on many parameters such as weather, soil, seed quality and farm practices. Importance of different parameters on crop is varying from crop to crop and region to region. With the availability of satellite images along with statistical data, deep leaning based model can capture growth of crop over temporal data. In this paper, we introduce informal methods to analyze and measure the impact of weather on the crop yield prediction using remote sensing images. Here, we applied Convolutional Neural Network (CNN) - Long Short-Term Memory (LSTM) based model over a large spatial and temporal data collected during crop growth season. We compared our model with two other models and found that it confers performance improvement over other models. From the experimental result, we dissect that inclusion of weather increase yield prediction and crop growth is highly correlated on the weather parameters.

Keywords: Crop yield · Weather · Analyze · Long short-term memory · Convolutional neural network

1 Introduction

For the many nations across the world, Agriculture is primary source of income. As per the Food and Agricultural Organization (FAO), approximately 10.5% of the total population of Asia suffers from the Food Insecurity [2]. India is one of the predominantly agrarian economies with more than 20 percentage shares of Agriculture and Allied Sectors in Gross Value Added (GVA) of the country (Records of National Statistical Office (NSO)) [1]. Around 54.6% of populations of India are associated with agriculture and its allied sectors [3]. In India, Agriculture depends on the rain for the water and other nutrition. Crop yield prediction is a complex and multifaceted task and it depends on many factors such as weather, soil, rainfall, farm practices etc. Majority of the farmers in India have small farmland (less than 2 acres) and accurate and reliable yield prediction can help them to improve production and better future planning. Weather plays an important role for the growth of yield. According to the Fifth Assessment Report (AR5) of the Intergovernmental Panel on Climate Change (IPCC), earth's surface temperature

© The Author(s), under exclusive license to Springer Nature Switzerland AG 2022
P. P. Roy et al. (Eds.): BDA 2022 India, LNCS 13773, pp. 249–259, 2022.
https://doi.org/10.1007/978-3-031-24094-2_17

is expected to increase over the twenty-first century and it highly affect heatwaves and precipitation as well as leads to increase its the intensity and duration in many parts of the world (IPCC, 2013b) [8]. It can lead to the decline in the global crop production.

Large human efforts are needed in the traditional manual survey methods for crop yield and it's not feasible for the country like India. Another alternative of the traditional methods of crop yield prediction is to perform yield prediction using statistical methods but these methods need large amount of soil data, weather data, farm practices related data, genetically related data at a finer level [4]. With the advancement in the image processing and space technology, remote sensing data also find application in the different domain such as weather prediction [5], rain fall and flood prediction [6], land usage land cover mapping [7].

The main aim of this work is to identify the impact of the weather parameters for the crop yield prediction using deep learning models. Growth of crop yield can be observed over the time series data hence we proposed Convolutional Neural Network (CNN) - Long Short-Term Memory (LSTM) based model. Instead of directly applying LSTM model to the time-series data, we first applied CNN to extract features automatically and then applied LSTM for capturing crop growth and impact of weather on it. The rest of the paper is divided into 5 sections which are as follows. Section 2 deals with the brief review of related work. Section 3 describes the dataset and proposed methods. Section 4 presents the result analysis of the proposed model. Finally, Sect. 5 presents the conclusion of the proposed method.

2 Related Work

In the past, Fisher have explored and identified the influence of weather on the yield prediction [9] and Baier has developed weather data based model for the yield prediction [10]. Apart from manual survey method, crop growth models are also used for the crop yield prediction and it perceive the plant growth based on the different plant parameters along with the environmental condition but it works moderately on finding historical incongruities on crop yield [11, 12] using historically data. In addition to crop growth models, statistical models also used as an alternative for finding the impact of weather on the crop yields [13, 14].

Prasad et al. [15] employ multi-variate regression based model to perform yield prediction using soil moisture, surface temperature, rainfall data and normalized difference vegetation index (NDVI) in the IOWA state of USA. Several machine learning and deep learning based algorithms such as Random Forest (RF), K-Nearest Neighbor (KNN), Support Vector Machine (SVM), Least Absolute Shrinkage and Selection Operator (LASSO), Ridge Regression, Regression Tree (RT), Deep Neural Network used for the crop yield prediction. Saeed et al. [16] developed deep neural network (DNN) based models for the crop yield prediction using genotype data and weather data. Kamir et al. [17] applied machine learning models on the rainfall and temperature data, MODIS data and observed yield map and identified yield gap in the wheat production.

Zipper et al. [18] identified that drought is one of the key reasons for the inconsistency of corn and soyabean crop production and observed spatial variability of the drought impact in the USA. Schauberger et al. [19] experimented crop growth models and found reduction of agricultural productivity due to high temperature. Kalcic et al. [20] had observed that weather positively influence increases of phosphorus loading which leads to the nutrient reduction of the soil in the Western Lake Erie Basin, USA. Yadav et al. [21] experimented that the inclusion of climate data with remote sensing data for the crop yield prediction increase the yield prediction for the crop wheat, sorghum, and corn at New Mexico.

Table 1. Summary of literature survey of research papers

Author	Important Parameters	Methods	Remarks
Zipper [18]	Precipitation deficit calculated from Meteorological Dataset	Regression based model	Identified fine-scale spatiotemporal patterns of drought effects on crop yield using meteorological data
Schauberger [19]	Minimum and Maximum Temperature, Precipitation	Statistical model	Observed negative response on crop growth due to high temperature
Yadav [21]	Minimum and Maximum Temperature, Vapor Pressure, NDVI	Linear, Polynomial, and Spline cubic regression model	Identified parameter influence the crop
Jiang [25]	Precipitation and Heat Stress	LSTM based model	LSTM based model learn general pattern from the spectral, spatial and temporal features
Sun [35]	Weather data, Remote Sensing Images	CNN-LSTM based model	Proposed model improved accuracy of yield prediction
Tian [36]	Precipitation, Average Temperature, Remotely Sensed Leaf Area Index (LAI) and Vegetation Temperature Condition Index (VTCI)	LSTM based model	Shows improved yield prediction and better adaptability to climate fluctuation
Sun [37]	MODIS surface reflectance data,	Multilevel CNN-LSTM based model	Observed that environmental data has relatively low influence for yield estimation at the end of the season

Weather and soil are important factors affecting the growth of crop and yield. From the literature survey, it is observed that weather plays vital role for the crop growth hence we aim to recognize the influence of weather on the crop yield prediction using deep learning based spatio-temporal model.

Here we used Convolutional Neural Network (CNN) - Long Short-Term Memory (LSTM) based models to capture spatial and temporal changes occurred during crop growth periods using the historical satellite images and weather and yield data. CNN algorithm is good for feature extraction and found application in different areas such as segmentation, classification and detection [22]. LSTM algorithm is good for capturing long-term temporal dependency using time series data in multivariate data. LSTM has shown significant results in various applications such as natural language processing [23], engineering systems [24] and corn yield estimation [25]. Summary of few papers is shown in Table-1. In this paper, we analyze the impact of weather on wheat as it is one of the cereals consumed as main food and supply diet for the necessity of calories and protein [26] and India is one of the largest wheat producers in the world [27].

3   Dataset and Methods

For this study, we built a model based on CNN-LSTM to identify pattern from the multisource spatial and temporal data. Instead of performing manual feature extraction, we prefer automatic feature extraction using CNN.

3.1   Study Area

The study area of this work is Gujarat state, which is one of the major wheat producing states of India. Gujarat is situated at Longitude 72.00 E and Latitude 23.00 N. Gujarat has rich diversified weather pattern from heavy rain fall area to the very low rain fall area like Kutch district. Apart from that, Gujarat has a large costal boundary which makes inimitable to identify weather impact compared to other wheat producing states. Duration of the wheat season is from mid of the September to the March. For the experimentation purpose, we had taken average yield production data of the years 2010–2020 shown in Table-2. Yield data is available in the public domain by the Crop Production Statistics Information System of Department of Agriculture Cooperation, Government of India [28].

3.2   MODIS Image Datasets

The proposed work uses Moderate Resolution Imaging Spectroradiometer (MODIS) images available in the public domain [29]. MODIS is a vital instrument on-board on Terra and Aqua satellites acquiring data in 36 spectral bands [29]. In this work, we have utilized following spectral band's data.

MOD09A1 [31]: It is a terra product also known as MODIS Surface Reflectance 8-Day L3 Global 500m. This product provides surface reflectance of terra MODIS bands 1 through 7 at 500m spatial resolution corrected for atmospheric condition such as aerosols, gases, Rayleigh scattering.

MYD11A2 [32]: It is an aqua product also known as Land Surface Temperature/Emissivity 8-Day L3 Global 1km. This product provides land surface temperature

Table 2. Average crop area and Wheat Yield in Gujarat, India

Year	Average area (million hectares)	Production (million tones)	Average yield (kgs/hectare)
2010–11	1.58	5.01	75.95
2011–12	1.35	4.07	78.26
2012–13	1.02	2.94	73.05
2013–14	1.44	4.60	76.81
2014–15	1.17	3.29	90.17
2015–16	0.85	2.31	85.63
2016–17	0.99	2.75	87.74
2017–18	1.05	3.10	96.56
2018–19	0.79	2.40	96.35
2019–20	1.39	4.55	100.59

and emissivity from the data of band 31 and 32 at a 1 km spatial resolution with 8 days temporal resolution.

MCD12Q1 [33]: It is a terra and aqua's combined land cover product also known as Land Cover Type Yearly L3 Global 500 m. This product provides land cover types at a 500 m spatial resolution with 12 months temporal resolution. It recognizes 17 different classes identified by IGBP (International Geosphere-Biosphere Programme) together with 11 natural vegetation classes, three human-altered classes, and three non-vegetated classes. From the given product, we identified vegetation area and used them in the training.

3.3 Weather Data

In this study, we used weather data downloaded from the OpenWeather portal [30]. This portal contains hourly weather data of Gujarat State from the 2010 to 2020. Hourly data is converted into the daily and weekly data. Attributes of the weather data are Average Temperature, Maximum and Minimum Temperature, Dew point, Pressure, Humidity and Wind Speed.

3.4 Proposed Method

In our study, we built model based on LSTM and Deep CNN [34] and it has three modules. In the first module, a sequence of images is given to CNN to extract features from it. Second module is LSTM, which identifies temporal relationship from the features extracted from the images during the crop growing days. Last module is fully connected module that predicts the crop yield. Here, we have given brief of the proposed models.

CNN-LSTM based proposed model is shown in Fig.-1. Input to the models is sequence of MODIS multispectral images I1, I2 .... I24 – collected during crop growing season of wheat for one year. The size of each image is 256 × 256 × b, where b is the number of bands. Another input is weather data of the study area. We train model using two different input sequences. In the first input sequence, we only provide sequence of images data. Each single image contains 7 bands from MOD09A1 product, 2 bands of land classification from MCD12Q1 product, 2 bands of MYD11A2 product and 2 bands of vegetation indices from MOD13A1 product hence total of 13 bands. In the second input sequence, we provide image data and weather data of the study area.

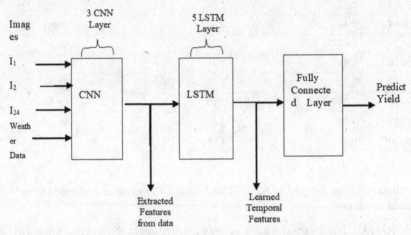

Fig. 1. Block diagram of the proposed approach

Figure-2 shows detailed architecture diagram of CNN-LSTM part of the proposed model. As shown in Fig.-2, CNN has three 3 layers, and each layer has convolution block followed by max pooling layer. First layer of CNN has 64 filters, second layer has 128 filters, third layer has 256 filters and size of each filter is 3×3. In our model, we have used max pooling as a pooling function and pooling size is of 5×5. ReLu is used as an activation function to add nonlinearity in the model. Output of the CNN is flattened into feature vectors, stacked according to the timestamp and passed to the LSTM layer.

In the first input sequence, only images provided to the CNN layer for the processing. During the second input sequence, images of size $256 \times 256 \times b$ where b is band is given along with weather parameters (Average Temperature, Maximum and Minimum Temperature, Dew point, Pressure, Humidity and Wind Speed). In the second input sequence, weather data also embedded along with the array of images data according to the time stamp of the images. It forms 3D tensor given to the CNN for the feature extraction and extracted feature is given to the LSTM layer.

Each LSTM layer has two parts; LSTM part followed by dropout layer with the dropout of 0.1 to learn temporal features. In the proposed layer, we have used simple LSTM model which is implemented using TensorFlow. To identify the loss of the model, we have used L2 loss function.

The quality of the extracted feature using CNN plays an important role for the better outcomes of the proposed model. Instead of directly giving images and weather data, we have given extracted features to the LSTM layer for learning temporal features of it. CNN is quite good in extracting spatial features, giving only selected features to learn temporal behavior helping the model to capture better results. The result of this model also depends upon the quality of the extracted features using CNN.

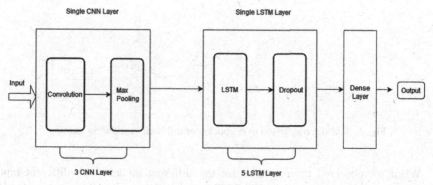

Fig. 2. Architecture diagram of the proposed approach

4 Result and Discussion

4.1 Model's Performance

We trained our model on the image and weather data of the years 2009–10 to 2018–19 and tested them on 2019–20 data. The model is trained for 30 epochs. We implemented our model in Tensorflow and trained using a GPU equipped with ×86 based Intel processor which uses NVIDIA Pascal architecture (P5000/6000). It has been observed that performance of a proposed model increased yield prediction with the inclusion of weather data. Performance comparison of different models has been shown in Table-3. In the Table-3, it shows that root mean square error (RMSE) of the model has been decreased more than 15%. It shows that the growth of yield is highly correlated on weather data. Any adverse change in the weather pattern can affect yield at a large extent.

Table 3. RMSE with and without weather data

Year	Without weather data	With weather data
CNN-LSTM	0.019261	0.002178
CNN-RNN	0.080132	0.0891
CNN-GRU	0.090545	0.0912

During the training phase, we increased the data of different years; RMSE is also decreased over the years. Figure-3 displays the comparison chart of four years of RMSE over the number of epochs and observed that error has been reduced with the increasing of data and epochs.

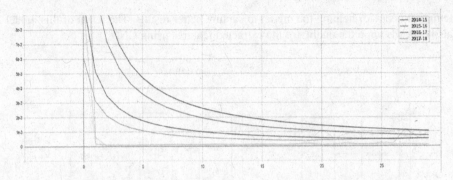

Fig. 3. RMSE comparison over epochs for different year during training

When we observed training time for the different models with different input sequences, we have noticed that CNN-LSTM model takes average 45 s per epoch with the images and weather data while same model takes average 38 s per epoch while training without weather data. Inclusion of the weather data increase the training time by 10 to 20 percentage compared to the without weather data. Average training time per epoch for the CNN-GRU models with image and weather data is approximately 43 s and without weather data it is 38 s. For the CNN-RNN model, it takes less training time per epochs compared to other models. Average training time per epoch for CNN-RNN model is approximately 40 s with image and weather data and 35 s without weather data. One of the observations is that the training time for the CNN- LSTM and CNN-GRU models are approximately similar in both the input sequences while CNN-RNN based model's training time is less compared to other models in all types of input sequences due to simple structure of RNN than GRU and LSTM.

4.2 Comparison with Other Models

We use root mean square error (RMSE) in kgs/hectare for comparing the performance of the different models. For comparison of a model, we also trained data on CNN-RNN and CNN-GRU based model. As LSTM is good for capturing temporal patterns, it has been observed from the Table-3 that our proposed model performs better than other models. RNN is good for handling small time series data but for handling large temporal data, LSTM is better choice due to effective gradient flow during backpropagation. Crop growth is a continuous growth and to capture pattern of growth over a period of time, temporal data is more feasible option. Another observation is that the inclusion of weather data also improves performance of the other models and reduces errors compared to without it.

5 Conclusion

In the given paper, we introduced deep learning-based model to analyze impact of weather on the crop yield prediction using publicly available satellite images and weather data. From the results, we can conclude that crop growth is highly impacted by the weather pattern. LSTM based proposed model can capture more weather impact than any other models over the temporal data. We also observed that with increasing the training data, error rate is also decreased and model learns features from the data. Here, we observed the impact of all the attributes of the weather data. In future, we plan to identify impact of different attributes of the weather data using other publicly available satellite images and make models interpretable.

References

1. PIB. https://www.pib.gov.in/PressReleasePage.aspx?PRID=1741942. Accessed 21 Apr 2022
2. FAO. https://www.fao.org/interactive/state-of-food-security-nutrition/en/. Accessed 22 Apr 2022
3. Annual Report 2020–21 Online Link. https://agricoop.nic.in/sites/default/files/Web%20copy%20of%20AR%20%28Eng%29_7.pdf. Accessed 22 Aug 2022
4. Bussay, A., Velde, M., Fumagalli, D., Seguini, L.: Improving operational maize yield forecasting in Hungary. Agric. Syst. **141**, 94–106 (2015). https://doi.org/10.1016/j.agsy.2015.10.001
5. Shin, J.Y., Kim, K., Ha, J.-C.: Seasonal forecasting of daily mean air temperatures using a coupled global climate model and machine learning algorithm for fieldscale agricultural management. Agric. For. Meteorol. **281**, 107858 (2020). https://doi.org/10.1016/j.agrformet.2019.107858
6. Karlsen, S.S.: Automated front detection-using computer vision and machine learning to explore a new direction in automated weather forecasting. The University of Bergen (2017)
7. Nguyen, H.T.T., Doan, T.M., Tomppo, E., McRoberts, R.E.: Land use/land cover mapping using multitemporal Sentinel-2 imagery and four classification methods—a case study from Dak Nong, Vietnam. Remote Sens. **12**(9), 1367 (2020)
8. IPCC (2013b). Summary for Policymakers, Book Section SPM, pp. 1–30. Cambridge University Press, Cambridge
9. Fisher, R.A.: The influence of rainfall on the yield of wheat at Rothamsted. Philos. Trans. R. Soc. Lond. Ser. B **213**, 89–142 (1925)
10. Baier W.: Crop weather models and their use in yield assessments. WMO Technical Note, p. 48, no. 151. World Meteorological Organization, Geneva (1977)
11. Müller, C., et al.: Global gridded crop model evaluation: benchmarking, skills, deficiencies and implications. Geosci. Model Dev. Discuss. **10**, 1403–1422 (2016). https://doi.org/10.5194/gmd-2016-207
12. Mistry, M.N., Wing, I.S., De Cian, E.: Simulated vs. empirical weather responsiveness of crop yields: US evidence and implications for the agricultural impacts of climate change. Environ. Res. Lett. **12**, 75007 (2017). https://doi.org/10.1088/1748-9326/aa788c
13. Vogel, M.M., Zscheischler, J., Wartenburger, R., Dee, D., Seneviratne, S.I.: Concurrent 2018 hot extremes across Northern Hemisphere due to human-induced climate change. Earth's Future **7**, 692–703 (2017)
14. Vogel, E., et al.: The effects of climate extremes on global agricultural yields. Environ. Res. Lett. **14**, 54010 (2019). https://doi.org/10.1088/1748-9326/ab154b

15. Prasad, A.K., Chai, L., Singh, R.P., Kafatos, M.: Crop yield estimation model for Iowa using remote sensing and surface parameters. Int. J. Appl. Earth Obs. Geoinf. **8**(1), 26–33 (2006)

16. Khaki, S., Wang, L.: Crop yield prediction using deep neural networks. Front. Plant Sci. **10**, 621 (2019). https://doi.org/10.3389/fpls.2019.00621

17. Kamir, E., Hochman, Z.: Estimating wheat yields in Australia using climate records, satellite image time series and machine learning methods. ISPRS J. Photogramm. Remote Sens. **160**, 124–135 (2022)

18. Zipper, S.C., Qiu, J., Kucharik, C.J.: Drought effects on US maize and soybean production: Spatiotemporal patterns and historical changes. Environ. Res. Lett. **11**, 094021 (2016)

19. Schauberger, B., et al.: Consistent negative response of US crops to high temperatures in observations and crop models. Nat. Commun. **8**, 13931 (2017)

20. Kalcic, M.M., Muenich, R.L., Basile, S., Steiner, A.L., Kirchhoff, C., Scavia, D.: Climate change and nutrient loading in the western Lake Erie basin: warming can counteract a wetter future. Environ. Sci. Technol. **53**, 7543–7550 (2019)

21. Yadav, K., Geli, H.M.E.: Prediction of Crop Yield for New Mexico based on climate and remote sensing data for the 1920–2019 period. Land **10**, 1389 (2021). https://doi.org/10.3390/land10121389

22. Li, W., Liu, K., Yan, L., Cheng, F., Lv, Y., Zhang, L.: FRD-CNN: Object detection based on small-scale convolutional neural networks and feature reuse. Sci. Rep. **9**(1), 16294 (2019)

23. Sutskever, I., Vinyals, O., Le, Q.V.: Sequence to sequence learning with neural networks. In: Advances in Neural Information Processing Systems, pp. 3104–3112 (2014)

24. Gangopadhyay, T., Locurto, A., Michael, J.B., Sarkar, S.: Deep learning algorithms for detecting combustion instabilities. In: Mukhopadhyay, A., Sen, S., Basu, D.N., Mondal, S. (eds.) Dynamics and Control of Energy Systems. EES, pp. 283–300. Springer, Singapore (2020). https://doi.org/10.1007/978-981-15-0536-2_13

25. Jiang, H., Hu, H., Zhong, R., Xu, J., Xu, J., Huang, J., et al.: A deep learning approach to conflating heterogeneous geospatial data for corn yield estimation: a case study of the US Corn Belt at the county level. Glob. Change Biol. **26**(3), 1754–1766 (2020). https://doi.org/10.1111/gcb.14885

26. Malik, D., Singh, D.: Dynamics of production, processing and export of wheat in India. J. Food Secur. **1**, 1–12 (2010)

27. Food and Agricultural Organization (F.A.O.). https://www.fao.org/india/fao-in-india/india-at-aglance/en/. Accessed 22 Apr 2022

28. Annual Yield Prediction Data Available. https://aps.dac.gov.in/APY/Index.htm. Accessed 22 May 2022

29. MODIS. https://modis.gsfc.nasa.gov/about/. Accessed 30 July 2022

30. OpenWeather Portal Page. https://openweathermap.org/. Accessed 16 Aug 2022

31. Vermote, E.: Mod09a1 modis/terra surface reflectance 8-day l3 global 500 m sin grid v006. In: NASA EOSDIS Land Processes DAAC. https://doi.org/10.5067/MODIS/MOD09A1.006

32. Wan, Z., Hook, S., Hulley, G.: Myd11a2 modis/aqua land surface temperature/emissivity 8-day l3 global 1 km sin grid v006. In: NASA EOSDIS Land Processes DAAC. https://doi.org/10.5067/MODIS/MYD11A2.006

33. Friedl, M., Sulla-Menashe, D.: Mcd12q1 modis/terra+aqua land cover type yearly l3 global 500 m sin grid v006. In: NASA EOSDIS Land Processes DAAC. https://doi.org/10.5067/MODIS/MCD12Q1.006

34. LeCun, Y., Bottou, L., Bengio, Y., Haffner, P.: Gradient-based learning applied to document recognition. Proc. IEEE **86**(11), 2278–2324 (1998). https://doi.org/10.1109/5.726791

35. Sun, J., Di, L., Sun, Z., Shen, Y., Lai, Z.: County-level soybean yield prediction using deep CNN-LSTM model. Sensors **19**(20), 4363 (2019). https://doi.org/10.3390/s19204363

36. Tian, H., Wang, P., Tansey, K., Zhang, J., Zhang, S., Li, H.: An LSTM neural network for improving wheat yield estimates by integrating remote sensing data and meteorological data in the Guanzhong plain, PR China. Agric. For. Meteorol. **310**, 108629 (2021) ISSN 0168–1923. https://doi.org/10.1016/j.agrformet.2021.108629

37. Sun, J., Lai, Z., Di, L., Sun, Z., Tao, J., Shen, Y.: Multilevel deep learning network for county-level corn yield estimation in the U.S. Corn Belt. IEEE J. Sel. Top. Appl. Earth Obs. Remote Sens. **13**, 5048–5060 (2020). https://doi.org/10.1109/JSTARS.2020.3019046

Analysis of Weather Condition Based Reuse Among Agromet Advisory: A Validation Study

Mamatha Alugubelly[1]([✉]), Krishna Reddy Polepalli[1], Anirban Mondal[2], S. G. Mahadevappa[3], Balaji Naik Banoth[3], and Sreenivas Gade[3]

[1] IIIT Hyderabad, Hyderabad, India
mamatha104@gmail.com
[2] Ashoka University, Sonipat, India
[3] P. J. Telangana State Agricultural University, Hyderabad, India

Abstract. India Meteorological Department (IMD) is delivering agromet advisories, i.e., weather-based crop risk management advisories based on the medium-range weather forecast (five days) across India. Based on the weather prediction, once in five days, agromet advisory is provided for major crops and livestock by considering the district/block as a unit. In the literature, a framework was proposed to improve the process of advisory preparation by employing the notion of *reuse*. In that framework, an approach was explored to reuse the advisory prepared for the given weather situation to prepare advisory for similar weather situations in the future. For this, a notion of category-based weather condition (CWC) was proposed to model a given weather situation. The experiments conducted by comparing CWCs of weather situations over a period of time showed a significant improvement in reuse. In this paper, we have conducted a validation study to analyze the scope of reuse by comparing the advisory text of the corresponding weather situations. The experiments on agromet advisory text data related to the Rice crop delivered from 2016 to 2019 for Telangana State show that if the advisory texts are similar, there is a high probability that the corresponding CWCs are also similar. The results validate that the CWC-based reuse framework can be employed to exploit reuse across weather situations.

Keywords: Data science · Reuse · Decision support systems · IT for agriculture · Agro-informatics · Crop management · Agrometeorology

1 Introduction

The weather influences many decisions in our day-to-day life as well as various activities in the domains of aviation, agriculture, business and healthcare. In India, weather forecasts are generated by the India Meteorological Department (IMD). To help the farming community, IMD has been delivering agromet advisories, i.e., weather-based crop risk management advisories based on the

P. P. Roy et al. (Eds.): BDA 2022 India, LNCS 13773, pp. 260–278, 2022.
https://doi.org/10.1007/978-3-031-24094-2_18

medium-range weather forecast (five days) across India [1]. The agromet advisory has been found to impact crop productivity significantly [7,12].

A weather situation is captured through the values of weather variables such as minimum temperature, maximum temperature, minimum relative humidity, maximum relative humidity and rainfall. Based on the weather prediction, once in five days, agromet advisory is provided across India for all crops by considering district/block as a unit. The past advisories can be reused if the current weather situation is similar to the past weather situation. There is a scope to improve the process of advisory preparation by exploiting the notion of *reuse*.

The notion of *reuse* in weather-based agro advisories has been exploited using the notion of *weather window* [4]. As an extension to this work, a case study to prepare the advisories for selected weather situations has been proposed in [14]. An attempt has been made to automate agro-advisory and provide a personalized crop-based recommendation to farmers [5]. In [10], a generalized framework to exploit the reuse by modeling weather situation with Category-based Weather Condition (CWC) was proposed. Efforts have also been made to explore reuse among weather situations at the block-level [9].

In the CWC-based framework [10], it has been considered that two weather situations are similar if the corresponding CWCs are similar. The intuition is that two similar CWCs should have a similar agromet advisory. The experiments were conducted based on 30 years (1986–2015) of weather data from the Rajendra Nagar weather station, Hyderabad, Telangana state, India. The results showed that there is a scope to improve reuse significantly by modeling weather situation with the notion CWC. The experimental results among the block level CWCs [9] have also showed significant improvement in reuse.

The goal of this work is to validate the CWC-based similarity among weather situations by comparing the corresponding advisory texts. Since the advisory is being delivered by agromet scientists, we consider that if the advice delivered for two different weather situations is similar, those weather situations should be similar. In this paper, by comparing the text advisory data, which has been delivered for weather situations, a validation study has been conducted to analyze the reuse. For this, the four years (2016 to 2019) of weather and the corresponding advisory data related to Rice crop delivered in Telangana state were used. The k-means clustering algorithm was employed to group advisories. The results are encouraging.

The key contributions of the paper are two-fold:

– We propose a validation framework to analyze the extent of advisory similarity among similar weather situations by employing CWC-based framework and the text analysis framework.
– We demonstrate the effectiveness of our CWC-based similarity framework by analyzing similarity among text advisories.

The remainder of the paper is organized as follows. In Sect. 2, we present the materials and methods. Section 3 explains the results and discussion. Finally, we conclude in Sect. 4 with directions for future work.

2   Materials and Methods

In this section, we first explain about the overview of agromet advisory service operated by IMD. Next, we briefly present CWC-based framework to exploit *reuse*. Subsequently, after presenting the methodology employed to compare the text advisories, we explain the experimental settings.

2.1   About Agromet Advisory Service

In June 2008, India Meteorological Department (IMD) started providing agromet advisories with 130 Agro meteorological Field Units (AMFUs) in collaboration with state agriculture universities and the Indian Council of Agricultural Research (ICAR). To improve the coverage, IMD has started providing crop-specific agromet advisories from 530 new District Agromet Units (DAMUs). In the future, IMD plans to deliver agromet advisories by considering block-level weather prediction, i.e., for each of 6500 blocks of India [1].

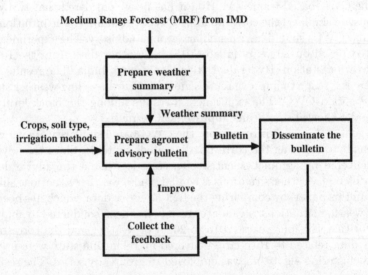

Fig. 1. Preparation and dissemination of agromet advisory

Figure 1 shows the steps for the preparation and dissemination of agromet advisory. Based on the predicted weather received for a given location/area, the domain expert at AMFU/DAMU prepares the agromet advisories to manage the weather-related risks for the major crops and lives stock of that location/area. The advisory is prepared twice a week. It is disseminated to farmers, and other stakeholders through the mKisan portal [3], Krishi Vigyan Kendra's [2], print, and electronic media. The feedback is being taken into account to prepare the subsequent advisory. Figure 2 shows a sample agromet advisory for the Rice crop from Rajendra Nagar. The first part of the advisory contains the weather information of a given location, and the second part presents the agromet advisory for major crops and livestock of the Rajendranagar area.

**PROFESSOR JAYASHANKAR TELANGANA STATE
AGRICULTURAL UNIVERSITY**
Agro Climate Research Centre (ACRC), Agricultural
Research Institute, Rajendranagar, Hyderabad-30.

Bulletin No. XXXIII/03/2022
WEATHER BASED AGROMET ADVISORIESFOR TELANGANA STATE
FROM DATE: 12.01.2022 (Wednesday) TO 16.01.2022 (Sunday)
Grameen Krishi Mausam Seva (GKMS Project)
INDIA METEOROLOGICAL DEPARTMENT, NEW DELHI.

Weather details of past 3 days:
During the last three days, dry weather prevailed over different parts of the State. The maximum and minimum temperatures ranged between 26-32°C and 15-21°C respectively.

Forecasted weather for the next 5 days:
As per the forecast received from Meteorological Centre, Hyderabad, light to moderate rain/thundershower very likely to occur at many places over Telangana during next five days. The maximum and minimum temperatures are likely to range between 26-33°C and 16-21°C respectively.

Warning:
Hailstorm and Heavy rains likely to occur at isolated places over Telangana.

Weather based Agromet Advisories

Rice:

- Raise rice nurseries of short duration varieties.

- Prevailing low temperatures may cause cold injury in rice nurseries. To overcome cold injury and for
  better nursery growth, adopt the following measures.

  ○ Cover the nursery beds with polythene sheet during night and remove inthe morning.

  ○ Irrigate the nursery bed every day in the evening and let out the water inthe morning.

  ○ Apply 2 kg urea for 200 m² nursery area at 10-15 days after sowing.

- Apply Carbofuran 3G @ 1 kg/200 sq.m (5 cents) to rice nurseries one weekbefore pulling nursery.

Maize:

● ...

Principal Scientist (Agro) & Head Agro Climate Research Centre (ACRC),
PJTSAU, Rajendranagar, Hyd-30.

Fig. 2. A sample agromet advisory

2.2   A CWC-based Reuse Framework

To improve the performance of the agromet expert who prepares the agromet advisories, a approach has been proposed in [10] by exploiting the notion of *reuse*. In [10], the weather situation of a location is captured with the concept of weather condition (WC).

Definition 1. *Weather Condition (WC)* *The weather condition for a set of n weather variables is denoted by* $WC_i = \langle l, d, s, n, V, \psi(V) \rangle$.

Here, i is the identifier of WC for the given location l and duration d, which is equal to the number of subsequent days, including the reference/start date s. The notation n represents the number of weather variables and V represents the set of n weather variables, i.e., $V = \{v_1, v_2, \ldots, v_n\}$. The function $\psi(\mathbf{V})$ represents the statistical summary of n weather variables such that $\psi(V) = \{\psi(v_1), \psi(v_2), \ldots, \psi(v_n)\}$. For each $v_k \in V$, $\psi(v_k)$ represents the statistical summary values of v_k computed for d days using the function $\psi()$. Normally $\psi()$ can be average, summation, or any other statistical function.

Table 1. Sample weather forecast from IMD

Id	Date	Tmax	Tmin	RHmax	RHmin	RF
Numerical values						
1	15/06/2018	35.5	23.5	80	48	0
2	16/06/2018	35.5	25.0	77	54	0
3	17/06/2018	36.5	24.5	81	47	0
4	18/06/2018	36.0	23.0	88	49	0
5	19/06/2018	34.5	22.0	88	64	4
6	20/06/2018	35.0	23.0	80	50	0
7	21/06/2018	37.0	24.0	77	46	0
$\psi(V)$		**35.7**	**23.5**	**81.57**	**51.1**	**4**
Categorical values						
1	15/06/2018	MF	LC	H	M	NR
2	16/06/2018	MF	AF	H	M	NR
3	17/06/2018	MF	LF	VH	M	NR
4	18/06/2018	MF	LC	VH	M	NR
5	19/06/2018	AF	LC	VH	H	LR
6	20/06/2018	MF	LC	H	M	NR
7	21/06/2018	MF	F	H	M	NR

Table 2. IMD's Weather Category Table (WCT)

Weather variable name	Range	Category name
Temperature (°C) (Deviation from Normal)	−1, 0, 1	Little change (LC)
	2 or −2	Rise/Fall (R/F)
	3 to 4	Appreciable rise (AR)
	−3 to −4	Appreciable fall (AF)
	5 to 6	Marked rise (MR)
	−5 to −6	Marked fall (MF)
	>7	Large rise (LR)
	≤−7	Large fall (LF)
Relative humidity (%)	0–30	Low (L)
	31–60	Moderate (M)
	61–80	High (H)
	≥81	Very high (VH)
Rain fall (mm)	0–0	No rain (NR)
	0.1–2.4	Very light rain (VLR)
	2.5–7.5	Light rain (LR)
	7.6–35.5	Moderate rain (MR)
	35.6–64.4	Rather heavy rain (RHR)
	64.5–124.4	Heavy rain (HR)
	124.4–244.4	Very heavy rain (VHR)
	≥244.5	Extremely heavy rain (EHR)

Table 1 contains one-week daily weather data from Rajendra Nagar weather station, Telangana, India. The units of $Tmax$ and $Tmin$, $RHmax$ and $RHmin$, and RF are degree centigrade (°C), per cent (%), millimeter (mm) respectively. The last row of Table 1 shows the WC for one week, i.e., $d = 7$ days. In Table 1, the summary statistics of the rainfall variable are computed by summing the rainfall values. Moreover, for the other variables, the average values are reported.

The issue, which was explored in [10], was that given a WC for n and d, say WC(n, d), for a given location and weather data of n variables, how do we compute similar WCs? The intuition is that if two WCs are similar, the agromet advisory prepared for one WC can be reused for the later WC. We can say that any two given WCs are similar, if the values of the corresponding weather variables are similar. Normally, if we consider numerical values of the weather variables and compare two WCs, it is difficult to find similar WCs. To improve the performance, the notion of category-based WC (CWC) was proposed. The

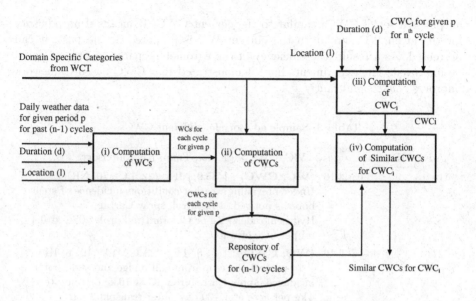

Fig. 3. Framework to exploit *reuse* among CWCs [10]

idea of CWC is as follows. In any domain, the value of weather variable is divided into categories based on the impact of the weather variable on that domain. The Weather Category Table (WCT) is defined as follows: For weather variable v_i and the set $C(v_i)$ of the corresponding weather categories, $WCT(v_i)$ is a two column table which represents the mapping of each value of v_i to the corresponding category in $C(v_i)$. Each row in $WCT(v_i)$ is of the form ⟨ range of v_i, category name ⟩. As an example, IMD has defined categories for weather variables temperature, humidity, and rainfall [11] (Table 2).

The notion of category-based WC (CWC) is defined by exploiting WCT. It is assumed that WCTs for all n variables are given. Given WC for n variables with duration d, WC(n, d), we can obtain the corresponding CWC(n, d) by replacing the each weather value with the corresponding category from WCT. Table 1 shows the WC and the corresponding CWC of one-week weather data for $Tmax$, $Tmin$, $RHmax$, $RHmin$ and RF.

Overview of CWC-Based Reuse Framework: Figure 3 depicts the CWC-based reuse framework [10]. Let n be the current cycle number for the given period p. Here, the notions of *cycle* and *period* indicate the number of consecutive days. A cycle is bigger than a period. For example, a year is an example of a cycle and a season is an example of a period. We assume that the values of daily weather data for V weather variables is available for $(n-1)$ cycles. Based on the domain requirements, the values of period p and duration d (in the number of days) are fixed. It is assumed that the WCT for the given weather variables for the given domain is available. We first compute the WCs and CWCs of the given period for $(n-1)$ cycles. For the current period, a CWC is given as input. The output is a set of CWCs similar to the current CWC. It means if an advisory for the given WC similar to the current WC is prepared in the past, it can be reused. As a result, the efficiency of the agromet preparation process could be improved. The experiments have demonstrated that CWC-based framework improves *reuse* significantly.

Table 3. Sample advisory for different CWCs

Advice ID	Date	<WC, CWC> {Tmax, Tmin, RF } advisory
1	18-03-2016	$\langle$**WC, CWC**$\rangle$: {$\langle$35.9, MF$\rangle$, $\langle$22, LC$\rangle$, $\langle$0, NR$\rangle$ } Under prevailing weather conditions incidence of stem borer is noticed. To control, spray Cartap Hydrochloride @ 2 g or Chlorantriniliprole 20SC @ 0.4 ml per litre of water
27	26-08-2016	$\langle$**WC, CWC**$\rangle$: {$\langle$31.8, TF$\rangle$, $\langle$23.1, MF$\rangle$, $\langle$104.8, HR$\rangle$ } To protect the crop from gall midge and stem borer incidence, apply Carbofuran 3G @ 10 kg or Phorate @ 4 kg per acre at 15-20 days after transplanting.
34	23-09-2016	$\langle$ **WC, CWC** $\rangle$: {$\langle$27.8, LC$\rangle$, $\langle$22.1, LF$\rangle$, $\langle$125, HR$\rangle$} Incidence of blast is noticed. To control, spray Tricyclazole @ 0.6 g per litre of water. In heavy rainfall received areas to avoid the incidence and further spread of Bacterial Leaf Blight (BLB) temporarily postpone (5–7 days) the application of Nitrogen fertilizers. Incidence of stem borer and leaf folder are noticed. To control, spray Cartap Hydrochloride @ 2g per litre of water. Incidence of panicle mite and grain discoloration is noticed. To control, spray Profenophos @ 2 ml + Propiconazole @ 1 ml per litre of water.

2.3 Methodology

The methodology employed to validate the CWC-based reuse framework consists of two steps. First, we analyse the similarity among CWCs using CWC-based reuse framework. Second, we analyse similarity by comparing the given two text advisories by employing text-based similarity computation framework, which is employed in the area of information retrieval. These methods are explained as follows.

- **Analyzing the similarity among WCs:** We have employed the framework proposed in [10] to compute similarity among two CWCs. By employing weather category table given by IMD (Table 2), we assigned the categories for each weather parameter v_i in WC to form CWC.
- **Computing the similarity among advisories:** To compute the similarity among two advisory texts, we have followed similarity framework proposed in the information retrieval literature. We also employ a clustering method to

form the groups of similar advises. The input to clustering process is advisories data, i.e., text documents. The date-wise non-empty Rice crop advisories are extracted for each year from 2016 to 2019 and assigned a unique advisory ID as shown in Table 3. As we computed experiments on Rice crop advisory only, we first extracted date-wise Rice crop advisories from the advisory bulletin. The total of 334 non-empty Rice crop advisories are considered as the corpus.

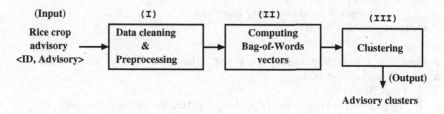

Fig. 4. Steps followed for advisory similarity

The following steps are followed to compute the clusters: Data cleaning and pre-processing, computation of Bag-of-Words vectors, and clustering (Fig. 4). We elaborate these steps.

(I) Data Cleaning and Pre-processing: The advisory data in the corpus is cleaned to remove the bullets and date-wise Rice crop advisory is listed and saved to a csv file for the pre-processing. The steps followed for pre-processing are as follows.

- **Removal of white-spaces and punctuations:** The extracted English advisory is converted into lowercase, and all white spaces and punctuations are removed from the advisory.
- **Remove stop words:** Stop words do not contribute any meaning to the sentence, so they are removed from the sentence. The NLTK (Natural Language Tool kit) library has a set of stop words, and we can use these to remove stop words from our text and return a list of word tokens. The stop words were removed by using the NLTK library.
- **Stemming and token generation:** Stemming is the process of getting the root form of a word. Stem or root is the part to which inflectional affixes (-ed, -ize, -de, -s, etc.) were added. The stem of a word is created by removing the prefix or suffix of a word. So, stemming a word may not result in actual words. We performed stemming and generated tokens using NLTK porter stemmer and word tokenizer methods.

(II) Computing Bag-of-Words (BoW) Vectors: The input to this step is the stemmed word tokens from the preceding step. The word-based feature representation method called Bag-Of-Words (BoW) [15] is used for advisory validation. A Bag-of-Words (BoW) representation of text describes the occurrence of words within a document. The tokens for each advisory are generated, and vocabulary is built by identifying the unique words in the corpus. Once the vocabulary is built, vectors are formed for each advisory by counting how many times each word appears.

Algorithm 1: Initclusters(adv,T)

Input   : adv[ ]: Array of n advisory vectors;
            T: User-defined similarity threshold
Output: Initial Clusters.

1 clusterID = 0;
2 Clusters= { }; // Dictionary to store cluserID: list of advisories
3 **while** *(size[adv]!= 0)* **do**
4 | tempCluster=[ ];
5 | tempCluster.add(adv[0]); // adding first adv to tempCluster
6 | **foreach** *tempAdv in adv[1:n-1]* **do**
7 | | **if** *cosine_Similarity(adv[0],tempAdv) >= T* **then**
8 | | | tempCluster.add(tempAdv);
9 | Clusters.add(clusterID, tempCluster) // Store advisory in tempCluster
   | to Clusters
10 | adv = adv - tempCluster
11 | clusterID = clusterID + 1;
12 return Clusters ;

(III) Clustering: To compare two vectors, we employ cosine similarity criteria [13]. Next, we present the clustering algorithm.

Cosine Similarity. To measure the similarity among two vectors, we used the cosine similarity metric. It is the cosine of the angle between two n-dimensional vectors in an n-dimensional space. It is the dot product of the two vectors divided by the product of the two vectors lengths (or magnitudes). The cosine similarity value ranges from 0 to +1, where 0 is dissimilar, and +1 is perfectly similar.

$$Cosine\_Similarity(A, B) = \frac{A.B}{\|A\| * \|B\|} \tag{1}$$

Here, A and B represents the advisory vectors generated by BoW model.

Clustering Method: To group the similar advisories into one cluster, we applied the clustering algorithm "k-means" [6] on our advisory data corpus. The k-means clustering is an unsupervised machine learning method, where given data points are grouped into k clusters. Normally, once the user defines the k value, k points are randomly chosen as initial centroids. We created the initial clusters using the approach presented in step 1 and then applied the k-means algorithm to our initial clusters. The step1 is as follows.

- **Step 1: Create initial clusters:** Initially, all the advisory vectors are formed using the BoW method, and the initial clusters are formed by computing the cosine similarity of the first advisory vector with all other advisory vectors in the corpus. This steps forms a initial set of advisory clusters with the given similarity threshold.

  The pseudo-code for this process is given in Algorithm 1. The input to this algorithm is advisory corpus (adv), user-defined similarity threshold (T). We initialize the first clusterID and a dictionary of Clusters to store clusterID and list of advisories (lines 1 to 2). The initial advisory (adv[0] is stored in the tempCluster (line 5). The similarity value of each adv in tempAdv and adv[0] is checked with the user-given threshold (T), i.e., cosine_Similarity(adv[0], tempAdv) $\geq$ T, where tempAdv = adv[i], i $\in$ 1 ... n $-$ 1 (lines 6 to 7), if yes then tempAdv is added to the tempCluster (line 8). Once we check for all advisories in the corpus, we form the final Clusters and remove these advisories from the main advisory corpus (lines 9 to 10). Then we repeat the above step, starting with the initial advisory in the remaining advisory corpus. The process is repeated till no advisory remains in the corpus.

- **Step 2: Find the final clusters through k-means clustering:** We now apply the k-means algorithm [6] on the initial clusters formed in step 1 to re-cluster the advisory by considering the first advisory in each cluster as the centroid. A k-means clustering is an iterative algorithm that groups a dataset into k pre-defined clusters. Each data point in the given dataset will belong to only one cluster.

Table 4. Advice similarity experiment settings

Type	Description
Location	RajendraNagar, Hyderabad, Telangana
Duration	Daily weather data ($d = 1$)
Weather dataset	4 years of daily weather data from January 2016 to the December 2019 for weather variables V = {$Tmax$, $Tmin$, RF}
Advisory dataset	4 years of Rice crop advisory with 334 advisories from the April 2016 to the December 2019
Weather CATEGORIES	WCT given by IMD
Methods	1. Bag-of-Words (BOW)
Similarity measure	Cosine similarity
Clustering algorithm	k-means
Experiments	1. Computing clusters of CWCs
	2. Computing clusters of advisory data

2.4   Experimental Setup

The experiments are conducted by considering the weather data, advisory data, weather categories defined by IMD. The experimental settings to present our methodology for computing the advisory similarity among similar weather conditions are presented in Table 4. We now explain the details of each parameter used in the experiment.

- **Location:** The experiments are conducted on the weather data collected from the Rajendra Nagar weather station, Hyderabad, Telangana state, India.
- **Duration:** Based on the requirement, weather variable values are computed for different duration's. These values are collected at different granularity levels like hourly, daily, weekly, monthly, seasons, etc. In this experiment, we considered weather data of five-day duration ($d = 5$).
- **Weather Dataset:** The dataset for the WC *reuse* experiment consists of daily weather data ($d = 1$) for four years, i.e., from the year 2016 to the year 2019, collected at Rajendra Nagar weather station, Hyderabad, India. The data consists of three weather parameters: $Tmax$, $Tmin$, and RF values, as major crop decisions, are made based on these three parameters. Agro-advisories are provided by examining five-day weather data. For this experiments, we have computed five-day WC ($d = 5$) from daily WCs ($d = 1$).
- **Advisory Dataset:** The dataset consists of agromet advisory for the Rice crop from 2016 to 2019. The total number of advisories for years: 2016, 2017, 2018, and 2019 are 64, 104, 103, and 99, respectively (Total = 370). The advisory is not available for the months of January and February in year 2016 and is also unavailable during the months of April, May, and November in each year as the crop reaches the maturity stage by those months. So, the advisory is not available for all WCs. Moreover, there are 36 empty advisories among these 370 advisories. The total non-empty advisories considered for the analysis are 334. The complete list of date-wise advisories can be accessed here [8].

  The sample advisory with advisory ID, date, WC, CWC, and rice crop advisories are presented in Table 3. These advisories are generated based on the five-day WCs with $d = 5$, once every five days. The advisory is given based on the crop problems that prevail for the given WC ⟨ Tmax, Tmin, RF ⟩ values. It can be observed that advisory is changed according to the WC values; for instance, from Table 3, we can observe the advisory with $ID = 1$ has no instruction related to rainfall as there is no rain (NR). For $ID = 27$, no instruction related to rainfall is observed though there is heavy rain as the crop has no harm with the rain in that stage. If we consider advisory with $ID = 34$, it has special instructions for heavy rainfall (HR). The advisory for prevailing pest are given based on the temperature and crop growth stage of the current location.
- **Weather Categories:** By employing the IMD's WCT [11] (Table 2, categories are assigned to each numerical weather parameter value of WC to form category-based WC (CWC).

- **Methods:** The Bag-of-Words (BoW) method is used to represent the feature vectors of the advisory.
- **Similarity Measure:** The cosine similarity metric is used as the similarity measure.
- **Clustering Algorithm:** We applied k-means algorithm to form the advisory clusters.
- **Experiments:** To analyze the advisory similarity among similar CWCs, we conducted two experiments, the first one is to compute the clusters of CWC using the framework proposed in [10]. The second one is to compute the advisory similarity using k-means clustering.

The results of the experiments are explained in the next section.

Table 5. Count of unique CWCs

No. of distinct CWCs (A)	Frequency (B)	Total CWC (A*B)
1	43	43
1	14	14
1	11	11
1	9	9
2	8	16
2	7	14
4	6	24
3	5	15
5	4	20
20	3	60
34	2	68
40	1	40
Total CWCs		**334**

3   Results and Discussion

This experiment is divided into two parts. In the first part, experiments were computed to extract CWC clusters by extracting the dates of occurrence of each unique frequent CWC. In the second part, we cluster the advisory using the k-means algorithm and map each advisory in the unique CWCs to the clusters formed using k-means.

Table 6. Advisory of top-five CWCs

CWC ⟨Tmax, Tmin, RF⟩ (frequency)	Advice IDs	Date of occurrence
CWC1 ⟨LR, LR, NR⟩ (43)	55, 56, 57, 58, 59, 60, 61, 62, 63, 64, 65, 67, 68, 70, 153, 154, 155, 156, 157, 158, 159, 160, 162, 163,164, 165, 167, 168, 169, 170, 171, 245, 246, 247, 248, 249, 250, 251, 252, 253, 255, 256, 257	27-12-2016, 30-12-2016, 03-01-2017, 06-01-2017, 10-01-2017, 12-01-2017, 17-01-2017, 20-01-2017, 24-01-2017, 27-01-2017, 31-01-2017, 07-02-2017, 10-02-2017, 17-02-2017, 15-12-2017, 19-12-2017, 22-12-2017, 26-12-2017, 29-12-2017, 02-01-2018, 05-01-2018, 09-01-2018, 19-01-2018, 23-01-2018, 25-01-2018, 30-01-2018, 06-02-2018, 09-02-2018, 13-02-2018, 16-02-2018, 20-02-2018, 24-12-2018, 28-12-2018, 02-01-2019, 04-01-2019, 08-01-2019, 11-01-2019, 18-01-2019, 22-01-2019, 25-01-2019, 01-02-2019, 05-02-2019, 08-02-2019
CWC2⟨AR, LR, NR⟩ (14)	47, 48, 71, 72, 73, 173, 174, 175, 237, 238, 260, 261, 262, 263	08-11-2016, 11-11-2016, 21-02-2017, 23-02-2017, 28-02-2017, 27-02-2018, 28-02-2018, 02-03-2018, 27-11-2018, 30-11-2018, 19-02-2019, 22-02-2019, 26-02-2019, 01-03-2019
CWC3 ⟨MR, LR, NR⟩ (11)	51, 54, 66, 69, 151, 152, 161, 166, 172, 236, 258	09-12-2016, 20-12-2016, 03-02-2017, 14-02-2017, 08-12-2017, 12-12-2017, 12-01-2018, 02-02-2018, 23-02-2018, 22-11-2018, 12-02-2019
CWC4⟨LC, LC, NR⟩ (9)	43, 44, 46, 228, 230, 231, 235, 299, 300	25-10-2016, 28-10-2016, 04-11-2016, 26-10-2018, 02-11-2018, 06-11-2018, 20-11-2018, 13-08-2019, 16-08-2019
CWC5⟨ LC, MF, RF-MR ⟩ (8)	120, 121, 123, 124, 139, 140, 315, 316	22-08-2017, 23-08-2017, 30-08-2017, 01-09-2017, 24-10-2017, 27-10-2017, 25-10-2019, 29-10-2019

3.1   Cluster Analysis of CWCs

The details of the CWC and advisory dataset used in this experiment are presented in Sect. 2.4. We consider $d = 1$ WCs of each advisory for the years: 2016, 2017, 2018, and 2019. First, we computed WCs for five-day duration ($d = 5$). Then, by using IMD's WCT [11], categories are assigned to numerical WC's to form CWCs. First, we computed the date of occurrence of repeating unique CWCs for 334 advisories covering the years 2015 to 2019. The number of distinct CWCs and the frequency is presented in Table 5. The top most frequent CWC repeated 43 times in four years. Similarly, the next three unique CWCs repeated 14, 11, and 9 times in four years. Further, we found 34 and 20 unique CWCs that occurred 2 and 3 times. There are 40 unique CWCs that occurred only once. From those, we picked the top- five CWCs (CWC1, CWC2, CWC3, CWC4, and CWC5) that occurred 43, 14, 11, 09, 08 times. A sample top-five frequent unique CWCs with respect to V = {Tmax, Tmin, RF}, dates of occurrence, and advice IDs are presented in Table 6. For instance, consider CWC1 that repeated 43 times (days) here we represent the date of occurrence and their corresponding advice ID, i.e., 27-12-2016 (55), 08-02-2019 (257). Similarly, for other CWCs (CWC2, CWC3, CWC4, CWC5), advice IDs and date of occurrence is presented in Table 6.

3.2   Cluster Analysis of Advisory Data

First, we considered all non-empty (334) advisories in the advisory corpus, and by applying Algorithm 1 on the corpus, we computed the initial clusters for the cosine similarity threshold (T) greater than or equal to 0.5, 0.6, 0.7, and 0.8, and obtained 14, 18, 26, and 41 clusters respectively. In the next step, advisories are clustered using the k-means algorithm. The k-means algorithm is applied by considering the first advice as centroid in each of the clusters with $k = 14, 18, 26$, and 41 $(T \geq 0.5, 0.6, 0.7,$ and 0.8). Then, we applied the elbow method to find the optimal value of k, which showed $k = 7, 9, 10$, and 20 as the optimal value. We re-clustered the initial clusters using the k-means algorithm by applying $k = 7, 9, 10$, and 20 for initial $k = 14, 18, 26$, and 41, respectively. The advice IDs for the clusters formed for cosine similarity $T \geq 0.5$ is presented in Table 8. Here, cluster ID, count of advisories in each cluster, and the advice IDs are presented. For instance for $T \geq 0.5$, the cluster with ID-6 has nine advisories with advisory IDs 87, 88, 89, 90, 91, 92, 93, 94, and 308. Similarly the advisory IDs of clusters 0 to 5 can be seen in Table 8.

Table 7. Cluster-wise count of advisories for top-five CWCs for T ≥ 0.5

Cluster ID	Count	CWC1 (43)	CWC2 (14)	CWC3 (11)	CWC4 (9)	CWC5 (8)
0	72	0	3	1	4	7
1	26	0	0	0	0	0
2	38	0	0	0	2	0
3	39	27	0	5	3	0
4	68	5	0	2	0	0
5	82	11	11	3	0	1
6	9	0	0	0	0	0

Finally, clusters are assigned to each advisory in the top-five unique frequent CWCs. The final clusters for the top-five frequent CWCs (CWC1, CWC2, CWC3, CWC4, and CWC5) for threshold $T \geq 0.5$ are given in Table 7. Here, we represent the count of advisory allocated to each clusters which are formed using k-means approach. For instance, from Table 7 shows the CWC1 has total 43 advisory that are mapped to clusters with IDs 3, 4, and 5. Similarly for CWC2 the advisories are mapped to cluster IDs 0 and 5. The CWC1 advice ids mapping to the advisory clusters formed using k-means algorithm are presented in Fig. 5. Here, we can see there is no overlap in the cluster IDs, each advisory will belong to only cluster.

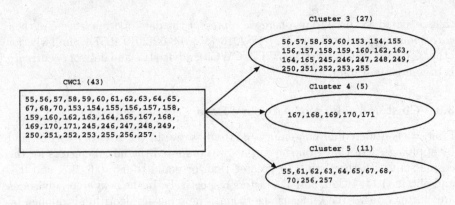

Fig. 5. Mapping CWC1 with advisory clusters for cosine sim >0.5

Table 8. Cluster for cosine sim T > 0.5

Cluster Id	Count	Advice IDs
0	72	34, 35, 36, 37, 38, 118, 119, 121, 122, 123, 124, 125, 129, 130, 131, 132, 133, 134, 135, 136, 137, 138, 139, 140, 141, 142, 143, 144, 145, 146, 147, 148, 175, 176, 177, 178, 179, 182, 211, 212, 213, 224, 225, 226, 227, 228, 229, 230, 231, 232, 233, 234, 235, 236, 237, 238, 239, 240, 266, 267, 268, 269, 313, 314, 315, 316, 317, 318, 319, 320, 321, 322
1	26	95, 96, 97, 98, 99, 100, 101, 102, 197, 198, 199, 200, 201, 289, 290, 291, 292, 293, 294, 295, 296, 297, 309, 310, 311, 312
2	38	6, 7, 8, 9, 10, 11, 12, 13, 14, 15, 16, 17, 18, 103, 104, 105, 106, 107, 108, 109, 110, 111, 202, 203, 204, 205, 206, 298, 299, 300, 301, 302, 303, 304, 305, 306, 307, 329
3	39	51, 52, 53, 54, 56, 57, 58, 59, 60, 149, 150, 151, 152, 153, 154, 155, 156, 157, 158, 159, 160, 161, 162, 163, 164, 165, 243, 244, 245, 246, 247, 248,249, 250, 251, 252, 253, 254, 255
4	68	0, 1, 2, 3, 4, 5, 19, 20, 21, 22, 23, 24, 25, 26, 27, 28, 29, 30, 31, 49, 50, 80, 81, 82, 83, 84, 85,86, 117, 166, 167, 168, 169, 170, 171, 172, 183, 184, 185, 186, 187, 188, 189, 190, 191, 192, 193, 194, 195, 196, 241, 242, 277, 278, 279, 280, 281, 282, 283, 284, 285, 286, 287, 288, 330, 331, 332, 333
5	82	32, 33, 39, 40, 41, 42, 43, 44, 45, 46, 47, 48, 55, 61, 62, 63, 64, 65, 66, 67, 68, 69, 70, 71, 72, 73, 74, 75, 76, 77, 78, 79, 112, 113, 114, 115, 116, 120, 126, 127, 128, 173, 174, 180, 181, 207, 208, 209, 210, 214, 215, 216, 217, 218, 219, 220, 221, 222, 223, 256, 257, 258, 259, 260, 261, 262, 263, 264, 265, 270, 271, 272, 273, 274, 275, 276, 323, 324, 325, 326, 327, 328
6	9	87, 88, 89, 90, 91, 92, 93, 94, 308

In the similar manner, we also computed the experiments with $T >= 0.6$, 0.7, and 0.8 and the advisory mapping count of each clusters formed using k means are presented in Tables 9, 10, and 11 respectively.

Table 9. Cluster-wise count of advisories for top-five CWCs for T > 0.6

Cluster ID	Count	CWC1 (43)	CWC2 (14)	CWC3 (11)	CWC4 (9)	CWC5 (8)
0	72	0	3	1	4	7
1	26	0	0	0	0	0
2	23	0	0	0	2	0
3	82	11	11	3	3	1
4	68	5	0	2	0	0
5	39	27	0	5	0	0
6	8	0	0	0	0	0
7	9	0	0	0	0	0
8	7	0	0	0	0	0

Table 10. Cluster-wise count of advisories for top-five CWCs for T > 0.7

Cluster ID	Count	CWC1 (43)	CWC2 (14)	CWC3 (11)	CWC4 (9)	CWC5 (8)
0	64	11	7	2	1	0
1	22	0	0	0	0	0
2	16	0	0	0	0	0
3	22	0	0	0	0	0
4	14	5	0	1	0	1
5	93	0	7	2	6	7
6	41	0	0	1	0	0
7	39	27	0	5	0	0
8	14	0	0	0	2	0
9	9	0	0	0	0	0

3.3 Discussion

From the results of advisory distribution among the clusters formed for $T \geq 0.5$ and $T \geq 0.6$ (Tables 7 and 9) the following observations can be made.

- The majority of the CWC1 advisories belong to a few clusters in each case, i.e., for $T \geq 0.5$ with $k = (14, 7)$, they belong to clusters 3, 4, and 5. From Table 9, for $T \geq 0.6$ with $k = (18, 9)$, they belong to clusters 3, 4 and 5 again. A similar trend is followed for the advisory mapping for $T \geq 0.7$ and $T \geq 0.8$ presented in Tables 10 and 11. From these results, it is clear that most of the advisories of similar CWCs are mapped to similar clusters, which show a high similarity among the advisories of each unique CWC.
- Also the list of date-wise advisories for unique CWC = {⟨Tmax, LR⟩, ⟨Tmin, LR⟩, and ⟨RF, NR⟩ are presented in Table 12. From this table, we can observe that the similar CWCs have similar advisory, i.e., the advisory for crop protection measures for low temperature was presented.

Table 11. Cluster-wise count of advisories for top-five CWCs for T > 0.8

Cluster ID	Count	CWC1 (43)	CWC2 (14)	CWC3 (11)	CWC4 (9)	CWC5 (8)
0	19	0	2	0	1	0
1	12	0	0	0	2	0
2	5	0	0	0	0	0
3	7	0	0	0	0	0
4	5	0	0	0	0	0
5	13	5	0	1	0	0
6	57	0	5	1	6	2
7	34	0	0	1	0	0
8	39	27	0	5	0	0
9	20	9	3	2	0	0
10	6	0	0	0	0	0
11	5	0	0	0	0	0
12	4	0	0	0	0	0
13	22	0	0	0	0	0
14	7	0	0	0	0	0
15	7	0	0	0	0	0
16	25	0	2	0	0	0
17	3	0	0	0	0	1
18	19	0	0	0	0	5
19	25	2	2	1	0	0

Overall, the above results show that similar weather conditions have similar advisory, so there is scope for reusing the agromet advisory if weather situations repeat in the future.

Table 12. Sample advisory for the Rice crop for CWC = LR/LR/RF-NR

Adv -isory ID	Date and advisory
56	**27-12-2016** Raise rice nurseries of short duration varieties (Kunaram Sannalu, Telangana Sona, Bathukamma etc.,). **Low temperatures** during the rabi may cause **cold injury** in Rice. To overcome cold injury and for better nursery growth, adopt the following measures: Cover the nursery beds with polythene sheet during the night and remove in the morning. Irrigate the nursery bed every day in the evening and let out the water in the morning. Apply 2 kg urea for 200 m^2 nursery area at 10–15 days after sowing. Apply Carbofuran 3G @ 1 kg/200 m^2 (5 cents) one week before pulling nursery.
57	**30-12-2016** Raise rice nurseries of short duration varieties (Kunaram Sannalu, Telangana Sona, Bathukamma etc.,). **Low temperatures** during the rabi may cause **cold injury** in Rice To overcome cold injury and for better nursery growth, adopt the following measures: Cover the nursery beds with polythene sheet during the night and remove in the morning. Irrigate the nursery bed every day in the evening and let out the water in the morning. Apply 2 kg urea for 200 m^2 nursery area at 10–15 days after sowing. Apply Carbofuran 3G @ 1 kg/200 m^2 (5 cents) one week before pulling nursery.
58	**03-01-2017** Incidence of stem borer and stem rot are noticed. To control, Stem Borer Apply Cartap Hydrochloride 4 G granules @ 8 kg or Chlorantraniliprole 0.4% granules @ 4 kg per acre. To control Stem Rot, Let-out water from the field and spray Hexaconazole @ 2 ml or Propiconazole @ 1 ml or Validamycin @ 2 ml per litre of water
246	**24-12-2018** Prevailing **low temperatures** may cause **cold injury** in rice nurseries. To overcome cold injury and for better nursery growth, adopt the following measures: Cover the nursery beds with polythene sheet during night and remove in the morning. Irrigate the nursery bed every day in the evening and let out the water in the morning. Apply 2 kg urea for 200 m^2 nursery area at 10–15 days after sowing. Apply Carbofuran 3G @ 1 kg/200 m^2 (5 cents) one week before pulling

4   Conclusion

In the literature, a category-based weather condition (CWC) framework was proposed to improve the process of agromet advisory preparation by employing the notion of reuse. In this paper, a validation study has been conducted to analyze the advisory similarity among similar CWCs using text-based similarity method. The results showed that the advisories of the similar CWCs mapped to similar clusters extracted based on text-based similarity to a considerable extent. Overall, the results are encouraging. As a part of future work, we are planning to investigate methods to improve reuse by exploiting latest advances in data science.

Acknowledgement. This work is supported by India-Japan Joint Research Laboratory Project entitled "Data Science based farming support system for sustainable crop production under climatic change (DSFS)", funded by Department of Science and Technology, India (DST) and Japan Science and Technology Agency (JST). We would also like to thank India Meteorological Department for providing the dataset for the experiments.

References

1. Standard operating procedure for agromet advisory services (2020). https://mausam.imd.gov.in/imd_latest/contents/pdf/gkms_sop.pdf. Accessed 30 Oct 2022
2. Krishi Vignan Kendra knowledge network (2022). https://kvk.icar.gov.in/agromet_advisory.aspx. Accessed 30 Oct 2022
3. A portal of Government of India for farmers welfare: mKisan portal (2022). https://mkisan.gov.in/. Accessed 30 Oct 2022
4. Balasubramanian, T., Jagannathan, R., Maragatham, N., Sathyamoorthi, K., Nagarajan, R.: Generation of weather windows to develop agro advisories for Tamil Nadu under automated weather forecast system. J. Agrometeorol. **16**(1), 60–68 (2014)
5. Dheebakaran, G., Panneerselvam, S., Geethalakshmi, V., Kokilavani, S.: Weather based automated agro advisories: an option to improve sustainability in farming under climate and weather vagaries. In: Venkatramanan, V., Shah, S., Prasad, R. (eds.) Global Climate Change and Environmental Policy, pp. 329–349. Springer, Singapore (2020). https://doi.org/10.1007/978-981-13-9570-3_11
6. Hartigan, J.A., Wong, M.A.: Algorithm as 136: a k-means clustering algorithm. J. Roy. Stat. Soc. Ser. C (Appl. Stat.) **28**(1), 100–108 (1979)
7. Maini, P., Rathore, L.: Economic impact assessment of the agrometeorological advisory service of India. Curr. Sci. **101**(10), 1296–1310 (2011)
8. Mamatha, A.: Github link for the Rice crop advisories (2022). https://github.com/mamatha104/weather-based-agro-advsiory-of-Rajendra-Nagar
9. Mamatha, A., Krishna Reddy, P., Balaji Naik, B., Sreenivas, G., Anirban, M., Seishi, N.: Improving efficiency of block-level agrometeorological advisory system by exploiting reuse: a study in Telangana. J. Agrometeorol. **23**(3), 330–339 (2021)
10. Mamatha, A., Krishna Reddy, P., Sreenivas, G., Seishi, N.: Analysis of similar weather conditions to improve reuse in weather-based decision support systems. Comput. Electron. Agric. **157**, 154–165 (2019)
11. Mazumdar, A.B., Medha, K.: Forecaster's guide, India Meteorological Department (2008). https://imdpune.gov.in/Weather/Reports/forecaster_guide.pdf. Accessed 30 Oct 2022
12. Rathore, L.: Weather information for sustainable agriculture in India. J. Agric. Phys. **13**(2), 89–105 (2013)
13. Singhal, A., et al.: Modern information retrieval: a brief overview. IEEE Data Eng. Bull. **24**(4), 35–43 (2001)
14. Balasubramanian, T.N., et al.: Designing agromet advisories for selected weather windows under automated weather based advisory system in Tamil Nadu - a case study. J. Agrometeorol. **18**, 34–40 (2016)
15. Wallach, H.M.: Topic modeling: beyond bag-of-words. In: Proceedings of the 23rd International Conference on Machine Learning, pp. 977–984 (2006)

Author Index

Alugubelly, Mamatha   260

Bamrah, Sumneet Kaur   117
Banoth, Balaji Naik   260
Belghith, Khaled   211
Bhardwaj, Mayank Ratan   227
Bhatnagar, Raj   172

Chaudhary, Sanjay   249
Chauhan, Arun   68
Chen, Siqi   172

Dadi, Kamalaker   16

Enaganti, Inavamsi   227

Fernandez, Ryan   149
Fournier-Viger, Philippe   185, 211

Gade, Sreenivas   260
Gayathri, K. S.   117

Haas, Roland E.   149
Haque, Rejwanul   91
He, Yulin   185

Jawadi, Jassem   211
Judy, M. V.   136

Kesavan, Ramesh   159
Kiran, Rage Uday   200
Kumar C., Pavan   68
Kumar, Ashutosh   104
Kumar, Shobhan   68
Kumaran, Kannan   91

Li, Yuechun   185
Likhitha, Palla   200

Mahadevappa, S. G.   260
Mondal, Anirban   260

Narahari, Y.   227
Nawaz, M. Saqib   185
Nisha Devi, K.   159

Parmar, Hasit   249
Patel, Pragneshkumar   249
Patel, Rohit   227
Pathak, Pramod   91
Peketi, Vijaya   53
Polepalli, Krishna Reddy   260

Raj, Tony D. S.   149
Rathore, Santosh Singh   104
Ravikumar, Penugonda   200
Roul, Rajendra Kumar   33

Saranya, R. B.   159
Satti, Surekha   53
Satyanath, Gaurav   33
Selg, Erwin   149
Sharman, Raj   3
Soman, Gayathri   136
Srivatsan, Shruti   117
Stynes, Paul   91
Surampudi, Bapi Raju   16

Talukder, Asoke K.   149

Vijayan, Bineetha   136
Vivek, M. V.   136

Waghmare, Abijeet V.   149
Watanobe, Yutaka   200

Printed in the United States
by Baker & Taylor Publisher Services